Driving Tomorrow

Our Roadmap to Sustainable Transportation, Infrastructure and Cities

Warren Gifford, PhD
David Turock, PhD

Copyright © 2019 Warren Gifford and David Turock
All rights reserved.

ISBN-13: 9781080850549

Table of Contents

Summary		v
Introduction		xi
1	Disruptive Innovations in Transportation	1
2	The Current State of Affairs	19
3	Designing the Ideal Transportation System	35
4	Autonomous Vehicles	51
5	Autonomous-Ways (A-Ways): A New Design for Infrastructure	72
6	Continuous Convoys	91
7	Autonomous Elevators	112
8	Linear Cities	128
9	C^4 ("See Forth") — Autonomous Navigation	145
10	Sustainability	160
11	Challenges from Big Business	179
12	Job Displacement Concerns	188
13	Social Challenges	198
14	Security and Privacy Protections	206
15	Political Challenges	219
16	Intentional Communities	228

17	Living Spaces	240
18	Learning Opportunities	251
19	Health Effects	265
20	Economic Drivers for an Inevitable Change	277

Conclusion   285

Summary

ONE TABLESPOON OF oil fuels a freight train to carry one pound of ripe oranges all the way across North America. Your car can't get out of the grocery store parking lot on one tablespoon of oil. In other words, the train is 30,400 times more efficient than your car. It's possible to shift our transportation system to be more like the train than the car. In this book we envision an autonomous transportation system in which this happens.

In our vision, vehicles will weigh less than their loads. A family of autonomous electric vehicles will circulate, delivering cargos as small as a single dose of medications or as large as 50 people. High-speed continuous convoys of vehicles will offer non-stop travel yet be 24 times more efficient with 15 times more capacity than our cars.

Enclosed routes will house and power these vehicles while doubling as enclosures for other forms of infrastructure such as water mains and electric lines. Cities laid out along these routes will combine the benefits of a city, the ambience of a town, and the green space of the country.

This new transportation system will be disruptive - compelling changes in parts of our society and economy that are currently frustratingly inert. New business opportunities and consumer demand will arise. The disruption will lead to more security and privacy, better learning, greater health, and a cleaner environment.

Does this sound like science fiction? In fact, all the technologies needed to realize this vision either exist or are on the drawing boards and within reach. In this book we show how such a transportation system would work and how you can be part of creating this future.

Introduction: A quick trip from Washington, DC to the Adirondacks, all while reading a novel, demonstrates some advantages of the new transportation system.

Section I – A New Beginning: Transportation for Tomorrow and Beyond

1. *Disruptive Innovations in Transportation*: Transportation has a long history of disruptive innovations from horses to wheels, trains, cars, and airplanes. What will the next innovation be?
2. *The Current State of Affairs*: Our current transportation system is expensive, inefficient, over-crowded, unhealthy, unsafe, outdated and deteriorating. It's time for a new disruptive improvement.
3. *Designing the Ideal Transportation System*: The ideal transportation system will have vehicles tailored to our needs, rather than bloated with excess weight. They will move us rapidly and safely with a minimum of time, effort, and resources.
4. *Autonomous Vehicles*: Autonomous vehicles come in different sizes to deliver our medications, beverages, foods, and anything else, just when we want them. Containers revolutionize freight transport so we can size them for our needs, with small ones nesting efficiently in larger ones.
5. *Autonomous-Ways: A New Design for Infrastructure*: What if we enclose our roads to keep out weather, debris, animals, pedestrians, and human-driven vehicles? These autonomous-ways, or A-Ways, come in different sizes and are also ideal to replace our aging pipes and wires.
6. *Continuous Convoys*: What if we could combine the convenience of local service with the speed of express? Continuous Convoys allow you to get on at any station and go full-speed to your destination. And they combine high speeds with amazing capacity and efficiency.

7. *Autonomous Elevators*: Elevators waste our time, use valuable floorspace, and consume a lot of power. Autonomous elevator vehicles could "drive" up and down a building giving us fast, efficient, private service.
8. *Linear Cities*: What if we build on top of the A-Ways? In addition to excellent transportation and infrastructure, they provide a physical foundation, and efficiently use the land currently wasted by roads.
9. *C^4 ("See Forth"): Autonomous Navigation*: Navigating safely and efficiently inside buildings around people and furniture calls for new approaches. Fortunately, technologies for online games and robots combine to create local cloudlets sharing information to solve these challenges with Cloudlet Computing, Communications, and Control.
10. *Sustainability*: Our current transportation system is only 20% energy efficient, yet electric autonomous vehicles drawing power from A-Ways are more than 95% efficient. Besides delivering our needs, they can whisk away our recyclables and wastes. Using the sun for electricity, heating, and light can make us fully sustainable.

Section II: Overcoming the Obstacles

11. *Challenges from Big Business*: Automobile and oil companies have been dragging their feet on sustainability. Yet Tesla has driven the universal shift to electric vehicles, and electric roofs. The universal mobility system opens doors for new business opportunities and improvements to existing businesses, which will overwhelm foot draggers.
12. *Job Displacement Concerns*: Debate rages over whether robots will eliminate jobs or create enough new ones. Universal mobility creates opportunities for new jobs as well as reducing the cost of living and giving us more time. Improved

education and health will smooth the transition to a better life for everyone.
13. *Social Challenges*: Millennials and others are already shifting away from our car-centric culture. Seniors are a rapidly growing segment of society, and as they see the many advantages of universal mobility, they will further accelerate this shift.
14. *Security and Privacy Protections*: Safety, security, and privacy problems flood the news. Autonomous vehicles and A-Ways are inherently safer than our current system. Innovative approaches to secure, private travel and deliveries offer comprehensive protection.
15. *Political Challenges*: Political forces challenge sustainable solutions. Yet the economic and social advantages of new systems, such as universal mobility, will build political pressures sufficient to drive change.

Section III – Transportation as the Core of a Sustainable Community

16. *Intentional Communities*: What if you could live in a community of supportive people with complementary interests and goals? Fast, inexpensive mobility overcomes the constraints of commuting to work and school. Matching neighbors and training for mutual support enable these intentional communities.
17. *Living Spaces*: The dense population of cities fosters creativity and economic advantage. Fast, inexpensive mobility can combine these benefits with the benefits of towns, and open spaces. Moveable walls and furniture increase the flexibility and efficiency of space enhancing these advantages.
18. *Learning Opportunities*: US education results are disturbingly poor. Fast, efficient transportation enables new learning approaches: move learners to opportunities matching their interests and needs; match seniors and youth for mutual learning and support. Intentional communities can provide

the supportive and stimulating environment essential for successful and enjoyable learning for all ages.
19. *Health Effects*: Time is critical in many health emergencies, so fast mobility saves lives and money. Most vehicle deaths and injuries will be eliminated. Access to health care and just-in-time medication can address the opioid and other health crises.
20. *Economic Drivers for an Inevitable Change:* The combination of desirability, cost savings and new business opportunities make change inevitable. Senior and health facilities using the new techniques will attract customers in preference to conventional facilities. Crowd funding, open source software and 3-D printers allow anyone to make small autonomous vehicles. Cities and airports can increase capacity and reduce costs, while providing better service for everyone.

Conclusion: Senior living, health facilities, airports, highway overlays, seaside communities, … compelling opportunities for initial applications of autonomous transportation. What part do you want to play in the excitement: innovator, builder, business owner, user, or roles yet to be invented?

Introduction

DESPITE BEING EARLY September, there was not even a hint of autumn in the air as David stepped outside to take his morning run. The hot and humid air hit him in the face as he welcomed the day outside his Washington, DC home, and before he got to the end of his street, he had decided it was time for a mini-vacation. His friend, Warren, had invited him to come stay for a few days after Labor Day in Lake George in the Adirondacks, and that sounded like the perfect place to make his escape. He hadn't seen Warren in a while, and the vibrant displays of fall colors amidst the cool, crisp air sounded absolutely wonderful.

After making sure Warren's invitation still stood, David packed a light bag for a weekend and searched the usual places for his smartphone. He finally found it near the last place he had been, and with phone in hand, he tapped on the transportation app to load in his destination, leaving as soon as possible. Within a few minutes, a personal mobility vehicle arrived to collect David with bag in hand. Of course, this was not some type of ride hailing service but instead part of the existing driverless, autonomous transportation system. Having ridden the system daily for several years now, David sat down as a matter of routine in an autonomous vehicle, which looked like a customized chair on a wheeled platform and stowed his bag under the seat.

Since David had already entered all of the information into the transportation system's application, his personal mobility vehicle proceeded to the nearest transportation access point. Though David's house was only two blocks away, even the farthest residents were within a mile or so of these access hubs. After reaching the hub, David's personal mobility vehicle entered a convoy vehicle, which almost immediately began to accelerate to match the speed of the approaching continuous convoy it would soon join. By accelerating to the same speed, it could simply join the front of the convoy and race along to its destination without stopping.

In the meantime, David was rather oblivious to the movements of his personal mobility vehicle as it moved from one speeding convoy vehicle to another. Instead, he was immersed in the last few chapters of the thrilling novel he so desperately wanted to finish. This might have been a rather simple task a few years ago when traveling over 400 miles, but this was no longer the case. In fact, the entire trip was going to take about an hour, much less than the 1 hour and 20-minute airplane flight plus over 2 hours to and from the airports, security screening and other delays. Regardless, David was confident he could complete the book before his arrival.

David's personal mobility vehicle seamlessly switched between convoy vehicles without any interruption. Since some destinations were more local or regional in nature, different convoy lanes moved at different speeds. But because David's destination was so far away, access to Elon Musk's hyperloop system was desirable. Once again, this required no effort on David's part as his personal mobility vehicle had automatically moved to a convoy vehicle that was making the same transfer. As the convoy approached the access hub to the hyperloop system, David's personal mobility vehicle entered a hyperloop transport and accelerated to achieve the 750-mph speed.

At first thought, you might be concerned about safety traveling at such a speed, but the current autonomous transportation system was much safer than any system previously used. Without human error as a variable, accidents had been nearly completely eliminated. Likewise, with all vehicles traveling inside a closed system of transportation lanes, other road hazards and weather concerns were non-existent. And naturally the vehicles all operated using electrical power supplied by solar panels located on the roof of the transportation structure, which not only made transportation energy efficient but environmentally safe as well.

David realized he had only finished three chapters of the book when he reached Albany, New York. In Albany, his vehicle exited the hyperloop and entered one of the convoy lanes that would take him the rest of the way to Lake George. David's vehicle then joined a convoy vehicle and completed his journey from Albany to Lake George. And once it had exited the transportation system access hub, David's personal mobility vehicle traveled independently the rest of the way to Warren's front door.

As David's personal mobility vehicle entered Warren's yard, David still had 2 chapters to read, so he checked his messages on his phone before tucking it away in his bag. The autumn air felt amazingly refreshing, and it was hard for David to believe he was sweltering in the Washington heat an hour before. David grabbed his bag from under his seat, stood up and knocked on the door. David's personal mobility vehicle then traveled to its next destination to provide transportation services to another individual close by.

"David! Glad you could make it," Warren exclaimed upon opening his front door. "How was your trip?'

"No complaints at all," David replied. "Hardly noticed I had left home until I saw the incredible fall foliage all around me."

If the preceding story sounds a bit like science fiction, it isn't. In fact, all the technologies required to make such a trip possible are already in existence or on the drawing board. Though the story only touches on a few of the features an autonomous transportation system could have, at a minimum, it highlights the remarkable advances we could soon enjoy in time efficiency, safety, and energy efficiency. In the remaining chapters, all of these concepts and many more will be discussed in detail. It is our hope that you will realize the tremendous potential that exists today for autonomous transportation systems, and with this realization, the wheels of change will gain speed toward this goal.

1
Disruptive Innovations in Transportation

HOW DID WE ever connect with one another before everyone had their own smartphone? If you are a millennial, you may not even recall such a time. Prior to this technology breakthrough, landline phones were used to reach someone, but this required them to be at home near the phone. If they were out to eat, in the yard, or down the street, you would simply have to try back later or leave a message on an answering machine. And if you wanted to find your way to a specific destination, forget GPS systems and Google maps…you actually had to unfold a paper map to get you there!

Smartphones and map apps have dramatically changed how we communicate with one another and find our way from place to place. However, their impact is much greater than this. Think about the role they play in the way we interact with one another. How many times do you see a couple on a date individually checking their messages and social media? New technologies also affect how we see the future. Would you have imagined drones delivering packages to your doorstep in the future if they were not hovering over us today? As each new technology appears, traditional ways of doing things suddenly shift. A new paradigm is introduced.

Often new technologies represent disruptive innovations. What is a disruptive innovation? Have you heard the phrase, "That's the best thing since sliced bread?" Believe it or not, sliced bread was a disruptive innovation. In the 1920s, bread makers resisted slicing their loaves of bread in fear that the bread would become dry. But for those companies that braved this new frontier, sliced bread rapidly outsold whole loaves. Why was this innovation disruptive? Few households needed toasters before sliced bread, and industrial bread slicers were non-existent. And while sandwiches were not new, the ability to quickly slap together a ham sandwich changed lunch options for many. In other words, sliced bread opened the door for new markets and changed how we lived.

Unlike incremental innovations, which result in gradual improvements in existing items, disruptive innovations result in major shifts leading to rapid change and progress. Making bread more nutritious is a great thing, but it is hardly disruptive. But slice the bread... wow! That is revolutionary! We know this seems a little silly, but it demonstrates the difference between disruptive innovations and incremental changes of an existing product. If something new dramatically changes the way we live our lives, then it is likely a disruptive innovation.

At its core, this book is about transportation, and disruptive innovations are hardly new to the transportation industry. From foot travel, to horseback, to trains, to cars, various innovations have suddenly changed how we get from one place to another. We believe the time has come for another radical shift in modern day transportation. New technologies offer exciting chances for us to greatly improve current systems. At the same time, many problems of current transportation methods can be resolved. And with these changes come wonderful opportunities for us to improve the quality of our lives. In this book, We invite you to explore these opportunities and envision what we believe will be the transportation for our future.

As a starting point, we take a look back at how transportation throughout history has evolved over time. We imagine it is difficult for you to describe yourself to someone else without delving into your past. When it comes to discussing the future of transportation, history is important as well. It shows us how new technologies changed entire societies (and even the world) for the better, and this, in turn, can show us the way forward today.

From Foot to Flight
How many steps have you walked today? Has your Fitbit reminded you to "step" things up a bit? It seems a bit ironic that amidst all the travel options we have today that the focus on foot traffic has once again become popular. Did you realize human foot travel is quite remarkable? Though we may not be the fastest creatures on earth, our endurance abilities of walking and running allowed our ancestors to overcome many obstacles. From long, seasonal treks to avoid nasty climates to successfully hunting (and sometimes avoiding) other animals, getting around by foot served us well for millennia.

Traveling by foot has natural limitations however. We can only cover so much distance in a period of time by foot, and we can only carry so much. If you have ever had to carry a loaded cooler from your car to the beach, this becomes evident rather quickly. This brings us to our first disruptive innovation…the ability to tame wild animals. Somewhere around 4500 BCE, several cultures began using animals to help get around. From horses and camels to donkeys and llamas, we began traveling and transporting goods with the aid of our four-legged friends. You may not consider this such a major development, but the use of animals for transportation greatly increased the distances traveled in a day and the amount of goods that could be carried.[1] After all Genghis Kahn founded the largest continuous empire

---

1 Casson, Lionel. *Travel in the ancient world.* JHU Press, 1994.

in history on horseback. The impact is still felt today. Don't believe it? Just consider the fact that we still measure cars in horsepower.

Can you guess the next disruptive innovation? Before you jump to something as glamorous as the automobile or train, don't forget about a more basic invention…the wheel. Images of B.C. comics may come to mind, but sometime between 3500 and 2500 BCE, a number of civilizations began using wheeled structures (with and without animals) for transportation and travel. Water travel also advanced around this time period from canoes and rafts to sailing vessels of various types. You can appreciate the changes that occurred with these advances. People from different areas could now trade goods, and groups were exposed to different cultures and beliefs. Often, these influences were positive, but they also resulted in wars, territorial disputes, and the pursuit of valuables.[2] You can see how transportation greatly affected many other areas of human life during this time.

The wheel was a big deal, and it was a long time before the next disruptive innovation in transportation occurred. After all, who could imagine anything more innovative than rounding off the corners of a cube! But things changed dramatically in the 19th century. With the discovery of the steam engine, railroads soon allowed continental transport, and newly constructed canals and steamboats offered the chance to reach many new places that were difficult to reach. Goods could be distributed among much larger areas, and people could travel longer distances in a shorter amount of time.[3] For example, you no longer had to live close to a dairy farm to get fresh milk, and coal mining flourished with the ability to transport coal.

With the Industrial Revolution, railroads, and steam engines, cities and civilizations changed greatly. While most major cities were already situated along waterways, railroads and steamboats resulted

---

2 Ibid.
3 Schivelbusch, W. (2014). *The railway journey: The industrialization of time and space in the nineteenth century.* Univ. of California Press.

in larger growth of these areas. The number of jobs increased in urban centers attracting more people to move from the country. At the same time, new cities began to appear along railroad lines with these increased travel options. Even commuter rails were developed allowing workers to live farther from where they worked.[4] With new ways to travel, civilizations again adapted to make the most of these systems.

While trains and steamboats greatly changed how we lived, they represented the tip of the transportation iceberg historically. Can you imagine life today without your car? We will get to this subject a little later, but your car is likely a valued part of your life. In the late 19th century, Karl Benz invented the automobile resulting in a discovery that would greatly impact societies throughout the world.[5] With the automobile, people could suddenly travel anywhere at any time for any reason all by themselves. Talk about freedom!

With this technology, trucks could suddenly haul materials long distances to areas waterways and railroads could not easily reach. Roads were constructed resulting in a web of connectivity never before seen. And before long, cars not only provided benefits in transportation but also a way to express one's individuality. Cars helped not only created a sense of personal independence and freedom, but they also became fashion statements and status symbols. Are you a Tesla person or a Mercedes type? The car you drive can say a great deal about your lifestyle and beliefs if you so choose. You can thus appreciate how the automobile greatly impacted our lives well beyond its ability to help us reach our destination.

Cars had a profound impact on city life as well. While jobs remained in urban areas, and commuter rails permitted suburban growth, cars expanded cities even more. Cars allowed less reliance on train schedules, and people could live in areas where commuter lines did not

---

[4] Ibid.
[5] Dietsche, K. H., & Kuhlgatz, D. (2015). History of the automobile. In *Gasoline Engine Management* (pp. 2-7). Springer Fachmedien Wiesbaden.

exist. Suburban neighborhoods began to pop up everywhere. More affordable homes and planned communities were appealing to many people, and the car allowed this to be a practical option. But with these changes came a few negatives. We are painfully aware of the traffic congestion, air pollution, and oil dependency these changes eventually brought. As you can see, disruptive innovations are not necessarily without costs despite the benefits they provide.

While the impact of the car on modern society was tremendous, it was not the last major disruptive innovation in transportation. At the turn of the 20th century, the Wright brothers made mankind's dream to fly a reality. Since then, flight travel has advanced from propeller planes to jets.[6] Not only can we travel across the country in a matter of hours via air travel, we can explore new continents and countries much more easily. Can you imagine visiting China before the introduction of air travel? The Queen Mary ocean liner took nearly four days to get from London to New York! Add on trains, automobiles and other boats to your journey, and it would likely take you two weeks to get there.

True globalization began with air travel, and at the same time, the American landscape changed again as well. Imagine what Memphis might be without FedEx. Or what about Louisville without UPS. Air travel disrupted previous delivery systems and enabled these companies to thrive influencing society and communities in major ways. Los Angeles became the major West coast hub in part due to the ability to travel by plane, and Alaska no longer had outrageous prices for everyday goods. And individual mobility increased leading to changes in personal lifestyles and job transfer practices. Because you could hop on a plane and be in another city within hours, the options of where one could live, or work expanded significantly. The traditional multigenerational family living together in a single city became much less common.

---

6 Kane, Robert M. *Air transportation*. Kendall Hunt, 2003.

Since the arrival of jet engines and the automobile, the transportation sector has not really experienced any major disruptive innovations. The last century has brought about a number of technological advances, especially in information and communication technologies, but these have not yet resulted in a major change to our daily commutes and travel. Certainly, our cars are equipped with backup cameras and an array of sensors to help avoid accidents, and not having Wi-Fi or a small television on your intercontinental flight seems just plain wrong in today's world. But we still drive our cars to work and catch a flight to travel abroad. Given the marked changes that occurred in transportation during the Industrial Age and shortly thereafter, it seems rather amazing that another major disruptive innovation has not emerged. We think that is about to change.

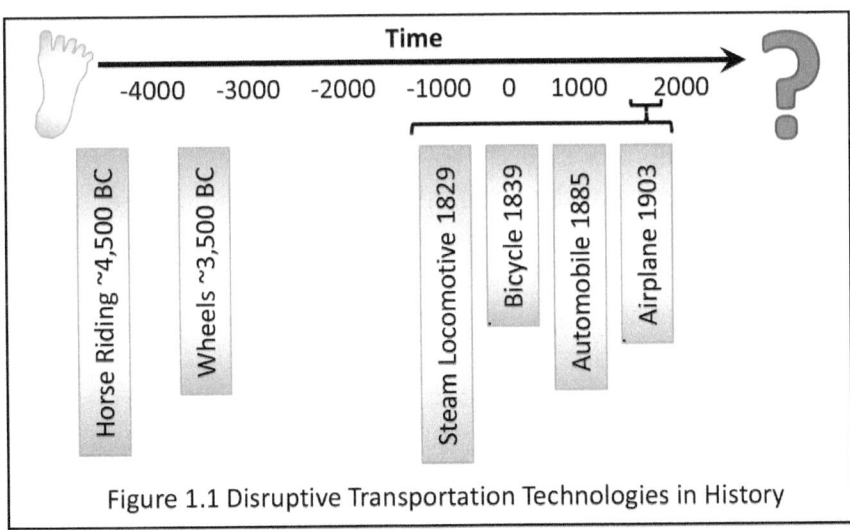

Figure 1.1 Disruptive Transportation Technologies in History

Transportation's Broad Reach

Warren:[7] *When I was a boy, my parents and I would travel by car to Florida to visit my grandparents for two weeks each summer in the*

---

7 When one of the authors relates a personal story we identify the source, and italicize the text.

1950's. *My parents would drag me out of bed at 2:30 am so that we could get through Washington, D.C. before the morning rush hour. Fortunately, I was allowed to bring my pillow and sleep in the back seat until a more reasonable hour! I remember the first summer trip took a very long time because of the single lane roads, dozens of slow trucks, and the difficulty the roads posed in the ability to pass slower vehicles. This added insult to injury after having to wake up so early.*

After that first summer, however, the drive each progressive summer took less and less time because of improvements in the roads. Also, each trip we would stay in a new motel (better than one the previous summer) with larger swimming pools and nicer rooms. These progressive improvements led me to believe that transportation and other services would continue to improve, presumably forever. And with the construction of I-95, which made the trip to Florida a breeze, my expectations were strengthened even more.

The interstate highway system was one of the greatest accomplishments in American transportation history. President Dwight D. Eisenhower made it a priority to pursue federal support for interstate travel because of the economic, social and commercial benefits it would bring to the nation. Interestingly, Eisenhower's first experience with interstate travel was in 1919. As a young lieutenant colonel in the army, he traveled as part of a convoy from Washington D.C. to San Francisco to assess the mobility of troops and materials in the event of an emergency. The convoy took 62 days and averaged 6 miles per hour![8] A marathoner might have made better time today!

The U.S. interstate highway system is certainly impressive. It provides almost 50,000 miles of limited access, high-speed transportation infrastructure that greatly changed the American landscape and culture. Seemingly overnight, destinations that previously took days of travel could be reached in a fraction of the time. Urban commutes from the suburbs also took less time resulting in the expansion of

---

[8] Pfeiffer, David A. "Ike's interstate at 50." National Archives, 2006. Retrieved from https://www.archives.gov/publications/prologue/2006/summer/interstates.html

city footprints. And new products could be more widely distributed to locations throughout the country as the trucking industry exploded.[9] In addition to advances in automotive technologies, these infrastructure changes resulted in dramatic changes in how we lived our lives in a relatively short time. You would be hard pressed to imagine our country without these highways today.

Of course, the interstate highway system was not necessarily favorable for everyone. In fact, it was a disruptive innovation itself. Gas stations, stores, and small towns along older state highways and county roads suddenly saw a marked drop in visitors. For some, this meant a substantial loss of business causing stores to close and residents to relocate. As some areas declined, however, urban and suburban areas grew substantially. Because the time to travel from city to city, or from suburban areas to city centers, was much less, homes and businesses became increasingly concentrated in metropolitan regions. Likewise, other businesses began to develop at entry-exit access sites along interstate roads. In other words, the design of the interstate infrastructure system had profound effects on where people lived and where businesses chose to establish themselves.

The interstate system offers a great way to see how transportation affects everything we do. It also provides the opportunity to see how one innovation can be both positive and negative depending on the setting. For decades, the federal government toyed with the idea of a national interstate highway system, but several attempts to develop such a system through state highway departments failed. One of the major developments that finally tipped the scales in favor of a national program was simply the number of vehicles on the road. In 1955, approximately 50 million cars and 10 million freight trucks used the existing network of roads and highways. Also, after World War II, the post-war economic boom tripled the number of freight

---

9 Ibid.

trucks needed throughout the country.[10] The demand for a better transportation system resulting from higher volumes of cars on the road and a need to move goods from one place to another.

Why did this growth trigger these demands for better interstate travel? For one, traffic and long commute times were becoming major deterrents in attracting workers to city businesses. Rural workers had little incentive to commute because of these delays. Likewise, the trucking industry pressed for change, so they could be more profitable. As far as the automobile was concerned, it was a disruptive innovation that progressively influenced infrastructure needs, and this in turn had profound effects on many other areas of society as we will highlight.

What does your car mean to you? In all likelihood, it represents a way for you to earn a living. Whether it provides an ability to commute from home to job, or whether it is a key part of your business, automobiles play a significant role in the ability to earn money. Parents also use their cars to take their children to and from soccer practice, the doctor, and any number of other places. Since children often have no other safe means to travel from one place to another, the car represents an important transportation mode for them. The same can be said for adult children with elderly parents who can no longer drive. These reflect the basic transportation values that cars award us on a daily basis.

Cars, however, offer much more than transportation. Your car may represent a private safe haven for you. Who hasn't hopped in their car, cranked up the sound system, and raced down the open road in order to feel that vivacious sense of freedom? In American culture, cars reflect our independence. In addition, your car may reflect your sense of fashion or personal achievements. Certainly, our cars allow us to show off our sense of style, and in many cases, it can help

---

10 Weingroff, Richard F. "Moving the goods: As the interstate era begins. U.S. Department of Transportation, 2017. Retrieved from https://www.fhwa.dot.gov/interstate/freight.cfm

define our level of success and status in society. In this regard, our cars reflect our individualism. From a personal perspective alone, the car influenced much more than our ability to get from one place to another.

Automobiles have also come to represent and influence other spheres of life as well. While our cars meet many of our transportation personal needs, automobiles have also come to own a large segment of the world sales and manufacturing markets. In fact, estimates suggest that automotive industry revenues, (including parts, repairs and re-sales) are over $4 trillion a year worldwide.[11] The automobile industry in general represents about 3.5 percent of the U.S. economy.[12] You can imagine how many jobs are associated with such an industry. The impact of the automobile on many nations' economies cannot be overestimated. However, it is also important to remember that the railroads once owned the transportation market before trucks and cars hit the scene ... they too had to eventually face change as painful as it was.

Every transportation system has the potential to affect key aspects of society, and these effects from the automobile are noteworthy. You may not even be aware of some of these effects. For example, consider health and safety. Thousands of deaths and millions of injuries occur each year in the U.S. from car crashes. Likewise, millions of dollars are spent on research and vehicle modifications in an attempt to address these issues despite the overwhelming majority being due to human error. Just think how texting while driving has increased the number of deaths on the road.

The car has created many social opportunities, but at the same time, it has also limited them. For example, disabilities, limited

---

[11] Scmitt, Bertel. "Auto industry 101: Today: How big?" Daily Knaban, 2015. Retrieved from http://dailykanban.com/2015/03/auto-industry-101-today-big/

[12] Industry and Trade Administration. "Automotive spotlight overview." SelectUSA.gov, n.d. Retrieved from https://www.selectusa.gov/automotive-industry-united-states

income, and age prevent many people from driving. Just think about the number of children, elderly and disabled who are not able to drive a car. This is a sizable percentage of society, and the car fails to effectively address these groups and their needs for transport and travel. Some mass transit and public transportation options exist, but these can be very restrictive and limited for these segments of the population.

Unfortunately, social, cultural and economic areas are not the only areas impacted by the automobile. In recent decades, the environmental impact of automobiles and their lack of sustainability as a form of transportation have become obvious. The harmful effects of carbon emissions, ozone layer depletion, noise pollution, lead residues in gasoline, and urban smog on the environment caused by automobiles are well recognized. Likewise, non-recyclable components of various automotive products and the footprint of roads and highways are other important issues affecting the planet. Did you know the paved areas in the U.S. occupy an area larger than the state of Georgia? And of course, our cars and trucks demand fossil fuels, obvious areas where sustainability is unachievable.

Based on this brief overview of automotive transportation, you can appreciate the widespread effects a new transportation system might have. Naturally, innovative transportation requires changes in infrastructure, which can have transformative effects also. The interstate highway system offers a nice example of this. But in addition, transportation affects many other areas that you may not immediately recognize. Urban designs, job opportunities, and economic impacts are routinely affected when disruptive technological advances in transportation occur. Likewise, changes in social dynamics, cultural norms, and societal trends typically change as well. And this fails to consider the profound and potentially devastating effects these systems can have on our environment.

Each of these areas will be explored in more detail in later chapters, but this overview allows us to appreciate the impact transportation can have in an ever-changing world. This appreciation also helps us place modern technological advances in transportation in a better perspective in relation to the future.

Today's Potential Disruptive Innovations in Transportation
If you recall, Karl Benz invented the car long before mass car transportation was available, and likewise, the Wright brothers' discovery predated commercial and military air travel by decades. Understandably, the technological advances that permitted true disruptive changes in transportation were created long before the actual changes took place. Applying this concept to our present-day situation, a number of potential disruptive innovations exist that could "rock" our world. Change typically comes about when both "push" and "pull" factors overcome the status quo. Where push factors are typically existing problems that require new solutions, pull factors represent more attractive options available. We believe a number of more attractive transportation options exist today.

You don't have to be a rocket scientist (or even a software engineer) to appreciate the revolutionary changes that have occurred in electronics, communications and other digital technologies. These changes are readily apparent from a generational perspective. For example, there is little doubt our great-grandfathers, if they were still living, could get into a new car today and figure out how to operate it. But give them iPhones, and they wouldn't have a clue what to do. A device many of us now take for granted was not even around before 2007! If such revolutionary changes can occur in the field of communication, then it certainly can take place in transportation... and in fact, it has.

Let's consider some of the gizmos and devices in our everyday world today. Did you know that the personal drone industry is a multi-million-dollar industry today? This is not referring to the unmanned military use of drones for surveillance and targeted attacks...this is simply consumer sales to private individuals and companies. The

FAA has registered over 1 million drones.[13] Even if you were aware of these figures, you may not be aware of the potential for these piloted machines to transport a variety of things from one place to another. Actually, this is not even a potential…it is already happening.

Drones have been in use to some extent since the 1990s, and they have been part of various military operations since the early 2000s. Drones have been used to transport military supplies and various forms of cargo. Beginning in 2011, K-MAX drone helicopters were used to transport millions of pounds of cargo to U.S. marines. These incredible machines are able to move 4,000 pounds of cargo at 15,000-foot elevations without having any crew whatsoever! And these drones have completed nearly 2,000 missions in total.[14]

In terms of commerce, Amazon has boasted that drone package delivery systems will be in place in a few years revolutionizing how consumers receive their merchandise (perhaps to the dismay of UPS, USPS and FedEx). Alphabet, the parent company of Google, has already experimented with drone delivery of Chipotle burritos on Virginia Tech's college campus using its "Project Wing" team in a FAA-approved testing environment. And Zipline, a private company, has been using drones in Rwanda to deliver medications, vaccines, and blood products in areas where transport limitations prevent easy access to these valuable supplies.[15]

While drones may play some role in specialized transportation services, air traffic issues, noise, inefficiency, and other things, will limit their use. However, the remarkable things about drones are their robotics and autonomous systems. Thirty-five years ago, autonomous mail delivery robots were being tested at the AT&T Long Lines Engineering Headquarters in Bedminster, New Jersey. When the cart

---

13 https://www.faa.gov/data_research/aviation/aerospace_forecasts/media/Unmanned_Aircraft_Systems.pdf

14 Symington, Steve. "Drones: The future of the transport industry is now." Newsweek, 2016. Retrieved from http://www.newsweek.com/drones-future-transport-industry-now-507080

15 Ibid.

would stop, it would beep, and people would come to collect their mail and deposit outgoing mail. The first (and only) time we saw the system in operation, the cart was bumping continuously into a wall. It soon became apparent that someone had replaced a carpet square containing the magnetic stripe that guided the mail robot in a different orientation. Instead of going straight down the corridor, it was trying to make its passage through sheetrock. Fortunately, the technology has come a long way in those 35 years and is racing ahead even faster with each year.

Let's look at a couple of other new technologies. iRobot began as a company in 1990 developing robots that were used to remove land and sea mines for the military. However, its current claim to fame is its autonomous robotic vacuum cleaner called Roomba. Since 2002, over 15 million of these robots have sold, and each one utilizes a variety of sensors to detect obstacles, dirty spots on the floor, and even steep drops so it does not fall down stairs.[16] Not only does it have the capacity to self-charge its own rechargeable battery, but it also actually goes farther on a kilowatt-hour of electricity than a Tesla Model S! While we love the Tesla, it doesn't vacuum our houses in the process of transport.

Another innovative technology in robotics is formerly known under the name of Kiva Systems. This powerful robotic warehousing system was so amazing in revolutionizing warehouse transport that Amazon chose to purchase the entire company for $775 million in 2012 and now has more than 80,000 robots roving warehouses. These compact (but extremely strong) robots are guided by barcoding on warehouse floors to find and transport shelving units containing a variety of products and materials. Based on an order's needs, the robot closest to the products required for the order is tasked with collecting the unit where the products are located. It then brings the unit to a person (known as a "picker") who assembles the actual

---

16 Kerr, Jolie. "The history of Roomba." Fortune, 2013. Retrieved from http://fortune.com/2013/11/29/the-history-of-the-roomba/

order for packing.[17] With built-in sensors to avoid running into each another, the Kiva robots perform an exquisite ballet gathering necessary materials quietly, efficiently, and accurately. And not a single person has to scour the warehouse to find needed materials in the process.

Other developments include the rise of sharing, ranging from vehicles, such as Zip Cars, to ride-hailing, such as Uber and Lyft, to shared bicycles. The rise of shared bicycles and the growing number of electric bicycles illustrates the shift to personal transportation vehicles. This shift is accentuated by the advent of even smaller powered personal vehicles, such as skateboards and scooters. These emerging trends highlight the growing desire to move away from our car-centric culture. The desire for personal freedom, convenience, and efficiency will continue to fuel the shift to new modes of transportation,

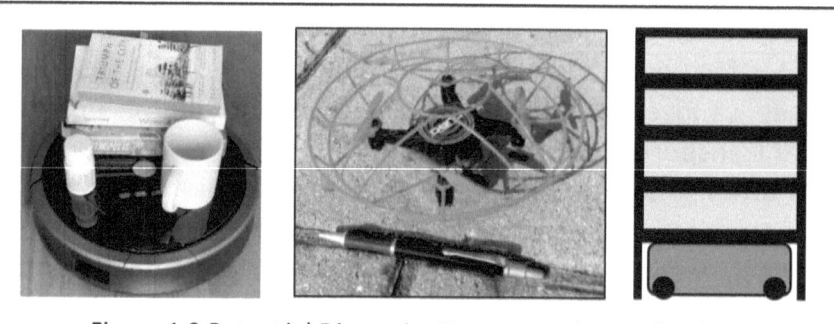

Figure 1.2 Potential Disruptive Transportation Technologies

This brings us to one of the most intriguing developments to date in transportation...the self-driving vehicle. In recent years, we have seen our cars come equipped with technologies that park our cars,

---

17 Tobe, Frank. "The technology gap left by Amazon's acquisition of Kiva Systems." The Robot Report, 2016. Retrieved from https://www.therobotreport.com/news/filling-the-void-left-by-kiva-systems-acquisition-by-amazon

automatically brake for obstacles, and guide us back into our lanes. These same sensors, computers and guidance systems are also being used to design driverless vehicles. Every automotive manufacturer on the planet, as well as companies such as Google, are investing serious resources in these concepts. Promises and predictions abound for when autonomous vehicles will arrive. A few fatal accidents have delayed introductions, but haven't diminished the determination to deliver.

While the initial driverless vehicles will likely involve driverless trucks traveling intercity routes, progressive advances are expected to occur in a relatively short time. Taxis, buses and other public transit options (in addition to trucks) are likely to be the first driverless automobiles, but in time, driverless consumer cars will appear offering us access to one of the most precious resources today…time. While the advances in these technologies are pulling us quickly into the future of transportation, the demand for time (whether it be less time in traffic or more free-time in transit — remember all that texting), is pushing us in that direction as well.

But even these opportunities to gain time pale in comparison to the innovations of Elon Musk. The Tesla Autopilot feature is already revolutionizing driving. In addition to driving all automakers to embrace electric vehicles, Musk is digging into new transportation territory with The Boring Company, tunneling under existing roads to make room for his high-speed innovation, the 750 mph Hyperloop. He is even thrusting transportation into space with the reusable SpaceX vehicles, noting in passing that using these vehicles you could travel between any two points on Earth in under an hour.

Disruptive innovations have repeatedly turned the transportation world upside down. In some instances, the need for a better transportation system accelerated the change, but in other instances, new inventions and discoveries provided new and better ways for us to get from one place to another. Today, we are again at a crossroads where new technologies and major transportation problems collide with the existing status quo. And with history as a guide, a disruptive change can not only be anticipated but expected. Everything is already in place for the perfect storm to occur, and given our current state of affairs, the changes to come should be very exciting. What part do you want to play in the excitement: innovator, builder, business owner, user, or roles yet to be invented?

2
The Current State of Affairs

NOT LONG AGO, the small town of Tupper Lake in upstate New York, needed to replace its storm sewers along the ¼-mile main street through town. Over the course of 3 months, dozens of workers and dozens of pieces of construction equipment dug down several feet, ripping out paving, pipes and wires, placing new conduit and manholes, and even leveling out a small hill, before replacing the paving, widening the sidewalks, and planting trees. Needless to say, this greatly disrupted traffic and the businesses on the street.

This seems like an extraordinary effort for such a small town. Yet the summer before, a similar effort disrupted a 200-yard stretch of road. Given the fact this stretch involved the larger of two roads by which people could exit Tupper Lake to the north, most residents and visitors alike were understandably frustrated with the delays. Likewise, the summer prior to that, the same work crews shut down a one-block section of road for a month to repair a water leak only to return a few months later to do it again!

These stories are not uncommon. If you live in areas of the country where snow frequently blankets the roads and temperatures freeze the ground, you appreciate the concentration of road repairs during the summer months. But the nation's transportation problem is not really one of weather or inefficient road crews. In reality, it's an issue of a failing infrastructure and outdated methods of repair in combination with outdated technologies. Have you noticed that even brand-new roads cause your car to bounce and thump along? To draw a comparison, it's like using and repairing old phone lines to run Internet connections. Does anyone remember dial-up Internet service? If so, then you can appreciate the level of frustration experienced by the drivers, businesses, and residents in Tupper Lake.

This chapter deals with our current state of affairs as it pertains to our transportation system. As you will see, infrastructure problems will be the common theme, but this involves more than just roads, bridges and sewer pipes. Costs, time delays, and safety issues are some of the major problems related to our poor transportation system infrastructure. Likewise, inefficient energy use, the transportation system footprint, and a failure to meet basic transportation needs are others. Having already talked about "pull" factors and disruptive technologies that promote change, this chapter will highlight the current problems that are pushing us toward change as well. Together, they make a strong case for a major paradigm shift in transportation.

The Costs of the Status Quo

Most of us drive our cars to work, to the store, and even out-of-town for vacation excursions. Some of us are chauffeurs for our children and parents. As a result, you probably feel you could render an accurate opinion about the current status of America's transportation infrastructure. Now, suppose you were a teacher and were going to give our nation's road systems, bridges, and airports a letter grade. What would it be? Would it pass or fail? According to the American Society of Civil Engineers, our country is not doing so well when it comes to transportation. In a comprehensive assessment of the U.S. infrastructure, they gave America a "D+!"[18] In other words, when it comes to our transportation structures, we are very close to failing.

---

18 ASCE. "2017 Infrastructure Report Card." InfrastructureReportCard.org, 2017. Retrieved from https://www.infrastructurereportcard.org/wp-content/uploads/2017/04/2017-IRC-Executive-Summary-FINAL-FINAL.pdf

To be fair, the definition of infrastructure used by the American Society of Civil Engineers extended beyond transportation alone. Infrastructure was defined as the framework provided by the country in meeting people's needs. It not only included roads, ports, rails, waterways, and airports, but it also included schools, energy systems, and waste-water management. Regardless, if we examine the individual grades for each of these areas that were assessed, the news is not any better. Roads, aviation, and public transit received a letter D grade while rails, bridges and ports were in the letter "C" to "C-" range. And inland waterways received the worst grade of all at "D-".[19] Hardly anything to feel warm and fuzzy about.

This brings us to our first area of concern regarding our transportation system, which is cost. In order to improve our nation's transportation infrastructure in its current state, the Federal Highway Administration estimates that $170 billion a year is needed.[20] The American Society of Civil Engineers estimates that a total of $3.9 trillion is needed to revamp all the areas of infrastructure by 2025.[21] By the way, these estimates are in addition to the dollars already budgeted for such improvements! Just think what else might be done with that kind of money...perhaps even a new and improved transportation system built from scratch? Let's not get too far ahead of ourselves just yet.

Clearly, transportation costs for repairs and maintenance are tremendous if not unsustainable. But believe it or not, these are not the only cost burdens related to transportation today. Did you know we personally spend over $1.2 trillion on transportation each year for our own individual use? After food, shelter and healthcare, people spend the highest amount of their income on transportation. On average, about 15 percent of your income is spent on transportation.

---

19 Ibid.
20 U.S. DOT, Federal Highway Administration. "2010 Status of the Nation's Highways, Bridges, and Transit: Conditions & Performance." Website, 2010. Retrieved from https://www.fhwa.dot.gov/policy/2010cpr/gr_highlights.cfm
21 ASCE, 2017.

And more than 90 percent of everything we spend for transportation is on our personal cars, fuel, maintenance, and car insurance.[22] Think about how much money you spend on your car the next time you're stuck in traffic or break down on the side of the road.

Transportation costs are also linked to the number of vehicles on the road. Did you know more than 250 million cars and trucks travel the U.S. highways and roads today? Estimates suggest that the number of automobiles in the world may exceed 2.5 billion by 2050 as nations like China "catch up" with other more developed countries in personal car use.[23] Since most cars now last over a decade, and since the production of new cars has barely slowed, this also means the number of clunkers accumulating in junk yards will grow as well. In fact, the automobile recycling industry is the 16th largest in the U.S.[24] You can appreciate how this quickly spills over into other problem areas like waste management, pollution, and energy shortages. In other words, the rising costs of one infrastructure segment can quickly result in rising costs in others.

When considering the transportation costs we experience, many different areas can be included. For now, we have focused on the financial costs related to a poor infrastructure and the burden we each pay to get from one place to another. But many other costs exist. Car crashes, traffic, carbon emissions, and excessive energy use are other important areas to consider. These will be highlighted in other sections but suffice it to say that the costs related to our

---

[22] Bureau of Transportation Statistics. Transportation Economic Trends. Washington D.C.: Bureau of Transportation Statistics Publications, 2016. Retrieved from https://www.rita.dot.gov/bts/sites/rita.dot.gov.bts/files/publications/transportation_economic_trends

[23] Tencer, Daniel. "Number Of Cars Worldwide Surpasses 1 Billion; Can The World Handle This Many Wheels?" Huffington Post Canada, 2013. Retrieved from http://www.huffingtonpost.ca/2011/08/23/car-population_n_934291.html

[24] LeBlanc, Rick. "Auto or car recycling facts and figures." The Balance.com. 2016. Retrieved from https://www.thebalance.com/auto-recycling-facts-and-figures-2877933

current transportation system are enormous. We truly cannot afford to maintain the status quo.

Energy Inefficiency at Its Worst

Let's compare the energy efficiency of our cars to that of a freight train. Did you know a freight train could bring a pound of ripe oranges from California all the way across the country on one tablespoon of oil? That's 30,400 times farther than your car can go on 1 tablespoon of oil. Seriously![25] We heard this story on the radio and promptly verified this claim with a friend who is a real railroad enthusiast. By comparison, the average car would not even make it an eighth of a mile on the same amount of fuel. In other words, you probably wouldn't be able to even get out of the grocery store parking lot! But what if you wanted the train to carry you as well as your oranges? In this case, the same tablespoon of oil would get you about 15 to 30 miles down the track. Even if you added your car to the train's cargo load, you would be able to travel about a mile on the same amount of oil, which is still eight times better than your car.

All things considered, our transportation system's energy efficiency is around 20 percent. In other words, our cars only effectively use about a fifth of the energy they require to run. By comparison, industrial energy efficiency is nearly 60 percent with commercial, residential and electricity energy efficiency all being higher than the transportation sector.[26] Unfortunately, more than a quarter of all of our nation's energy needs involve the transportation sector. As you can appreciate, much of the country's investments in energy go to waste simply to run a transportation system that uses energy quite poorly.

---

25 CSX. "Fuel efficiency." Website, 2016. Retrieved from https://www.csx.com/index.cfm/about-us/the-csx-advantage/fuel-efficiency/
26 Lawrence Livermore National Laboratory. https://flowcharts.llnl.gov/content/assets/images/energy/us/Energy_US_2016.png

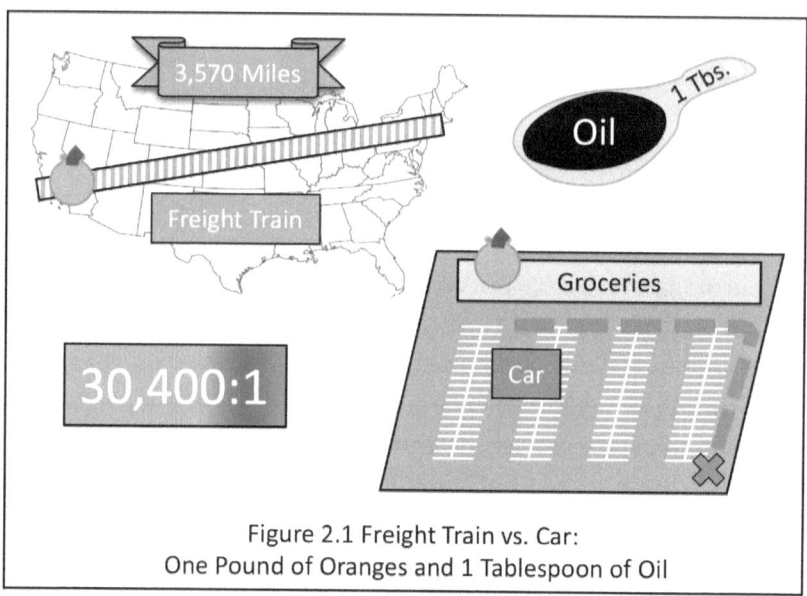

Figure 2.1 Freight Train vs. Car:
One Pound of Oranges and 1 Tablespoon of Oil

Certainly, electric cars have improved this situation to some degree, but the U.S. remains heavily dependent on oil. In 2016, we purchased over 10 million barrels of oil from foreign countries, and at about $50 per barrel, we currently spend nearly $500 million per year on imports![27,28] If cars could achieve higher energy efficiency, then foreign oil imports would not even be needed. Specifically, if fuel efficiency went from 21 percent to 53 percent, all foreign oil imports could be eliminated. Unfortunately, such improvements have been slow and insufficient in the transportation industry. This is a major reason why energy policies remain at the forefront of every major political election.

---

[27] U.S. Energy Information Administration. "How much petroleum does the U.S. import and export?" Website, 2017. Retrieved from https://www.eia.gov/tools/faqs/faq.php?id=727&t=6

[28] CrudeOilPrice.net. "Crude oil and commodities prices." Website, 2017. Retrieved from http://www.oil-price.net

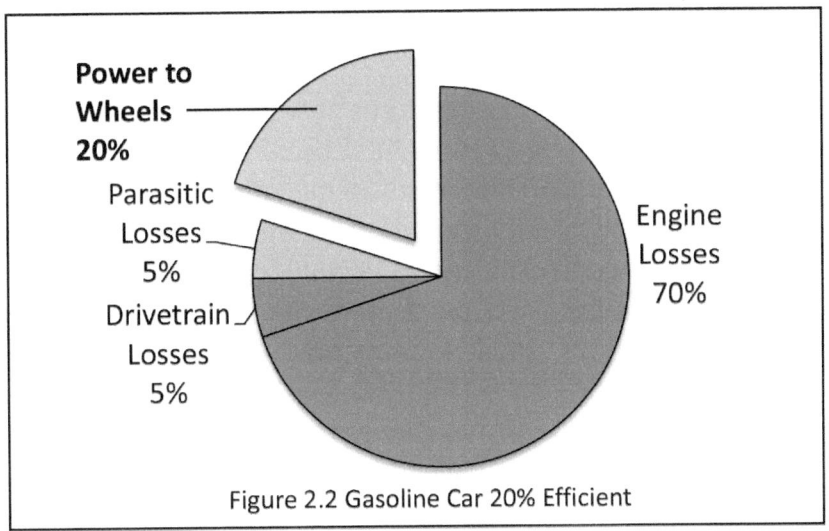

Figure 2.2 Gasoline Car 20% Efficient

Promoting Health or Injury?

If you have recently been to your doctor, you have likely heard buzzwords like prevention and health promotion. Unfortunately, we have a healthcare system designed to react to disease and injury instead of one designed to deter these hazards from the start. Nevertheless, we have finally come to realize healthy behaviors and early detection of problems offer not only better health but also much lower costs. The saying, "an ounce of prevention is worth a pound of cure," is actually true. And the same adage can be applied to transportation systems as well, especially when we think about our safety.

Did you know that car crashes have killed more people in the U.S. than all of the American wars combined? In fact, the numbers are not even close...motor vehicle crashes account for nearly three times as many deaths! The National Safety Council reports that car crashes caused about 40,000 deaths in the U.S. during 2016, and roughly another 4.6 million Americans suffer severe injuries from these events.[29] These are staggering numbers when you consider the goal is simply to travel from one place to another.

Car crashes account for significant costs from a number of different perspectives. Let's consider healthcare costs first. In the U.S. alone, car crashes result in more than 25 million emergency room visits and over 200,000 hospitalizations. With an average bill of $3,300 for these emergency room visits and $57,000 for each hospitalization, you can appreciate the costs associated with healthcare involved.[30] Now for the bad news...medical costs only account for 10 percent of the total cost associated with car crashes. Property damages, lost productivity from being injured, insurance costs, and legal expenses also contribute greatly to other direct costs. Even lost productivity when you are stuck in traffic behind an accident contributes to these losses.

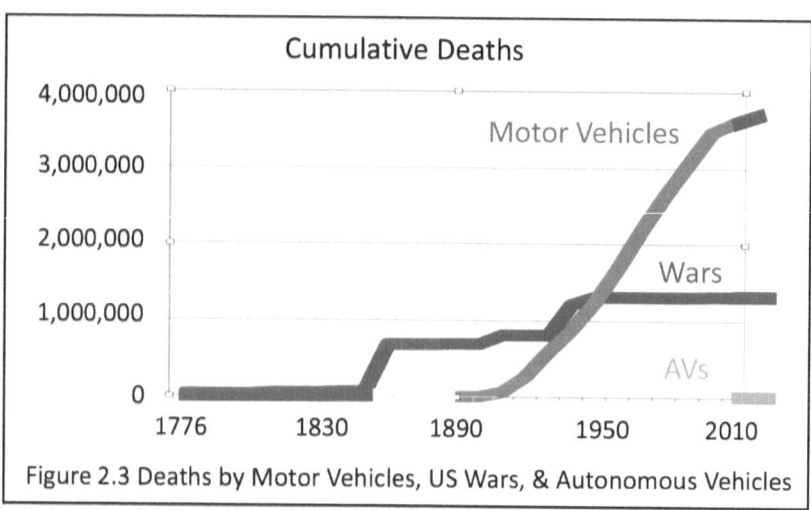

Figure 2.3 Deaths by Motor Vehicles, US Wars, & Autonomous Vehicles

---

29 National Safety Council. "Motor vehicle deaths in 2016 estimated to be highest in nine years." Website, 2017. Retrieved from http://www.nsc.org/Connect/NSCNewsReleases/Lists/Posts/Post.aspx?ID=180

30 Centers for Disease Control. "Motor vehicle crash injuries." CDC Website, 2014. Retrieved from http://www.cdc.gov/vitalsigns/crash-injuries/

In 2010, the sum of all of these related costs just described totaled $242 billion for the year. However, these costs did not consider the loss of quality of life resulting from injury or loss of a loved one. Though more difficult to measure, the National Highway Traffic Safety Administration calculates the number of quality life-years lost from car crashes, and while the other factors mentioned account for 29 percent of the cost, lost quality of life accounts for 71 percent! In other words, car crashes account for over $800 billion in costs each year.[31] While the National Highway Traffic Safety Administration seeks to make transportation safer, it would likely admit our current system of travel is far from ideal.

After many years of declining deaths from car crashes, recent statistics again show the number is rising. Texting and smartphone use may be a major part of the problem, but even without these hazards, the number of deaths and injuries from car crashes remain significant. Human error still accounts for more than 90 percent of all car crashes. Despite airbags and dozens of new "sensors" to help us, car crashes and personal safety remain major issues with our current transportation system.

A Timely Matter to Consider

Driving requires our attention. Sure, there are moments while driving down a barren stretch of interstate where it may seem like you are on auto-pilot, but even then, some level of attention is required. If nothing else, texting has highlighted this simple fact. Have you ever glanced at your phone screen for just a couple of seconds to check a message? Be honest. If you're like most drivers, you have. According to a major survey involving more than 3 million drivers and over 5.6 billion miles of travel, the results found that 88 percent of

---

31 National Highway Traffic Safety Administration. "The economic and societal impact of motor vehicle crashes, 2010 (revised)." U.S. Department of Transportation, 2015. Retrieved from https://crashstats.nhtsa.dot.gov/Api/Public/ViewPublication/812013

trips were associated with some phone use. Even worse, a mere two-second distraction was enough to increase the risk of a car crash by twenty-fold![32]

Certainly, these survey findings relate to our safety and health, but it also highlights another important finding...we all want more time. Just think if we did not have to drive to work, to soccer practice, to the airport. Many people who commute by subway or train appreciate the extra hours per week they have to read or do other things. The majority of us choose to drive instead, often out of convenience sacrificing time in the process. So, we attempt to "steal" some of this time back in tiny pieces as we drive thinking everything will be just fine. Unfortunately, sometimes it is not.

If you are like most Americans, you spend over 100 hours every year in your car. How about some more depressing facts? The average American spends over 40 hours a year in traffic, and traffic congestion costs us over $120 billion annually.[33] On a good day, we lose the time driving from one place to another, but when we encounter unexpected delays, breakdowns, and other "unavoidable" problems, the costs climb significantly.

If you go back in time and consider how long it took our forefathers to get from one place to another by horse, carriage, or railroad, then the time spent traveling today might seem relatively small. But a couple of points have to be made. First, with cars and planes, much of the time we might have saved was lost as we traveled longer distances from home. In other words, the time spent traveling stayed relatively constant while the distance to our destinations increased significantly. Secondly, time is a much more precious resource today when compared to decades past. Thanks to the explosion of

---

[32] Axios. "Study confirms we are all on our phones while driving." Axios.com, 2017. Retrieved from https://www.axios.com/study-confirms-we-all-use-our-phones-while-driving-2363176663.html

[33] U.S. Department of transportation. "Beyond traffic 2045: Trends and choices." DOT Website. Retrieved from https://cms.dot.gov/sites/dot.gov/files/docs/Draft_Beyond_Traffic_Framework.pdf

information and global access, time comes at a premium price, and time spent driving is often perceived as being a cost, not a value.

Expensive	Dangerous	Time Wasting
9% of Personal Consumption Expenditures $1.1 Trillion 5.6% GDP	4.6 million Injured/year. Cost of Crashes $826 Billion, 4.2% GDP	67 Minutes/Day @ $15/hour, Cost of Time $1.5 Trillion 7.6% GDP

Figure 2.4 Transportation Costs, Injuries and Wasted Time

All for One, But Not One for All

Warren: Many of us take driving a car for granted. I must say I did too until I saw my father have to give up his driver's license. Over several years, he began losing track of his whereabouts. The last time he drove, he seemed to control the car just fine, but every few minutes I found myself reminding him where to turn and where he was going. I drove the car back home that trip much to not only his dismay but to mine as well.

While my father was upset about giving up his driving privileges, my stepmother was even more distraught. Though their community provided free shuttle services around town and longer rides for a small fee, she loved the convenience and benefits their own car provided.

Certainly, inconveniences exist with public transportation. Transit schedules, riding with unknown passengers, and route delays have a hard time competing with the freedom and instant gratification of a personal car. But personal cars have significant limitations also in meeting our transportation needs. For my father and stepmother, these limitations became an undeniable reality just as it has for many other segments of the population.

Think about the limitations of our current transportation system in relation to age. If you're under 16 years of age, driving is not even an option. Children and adolescents represent a significant portion of the population, and because of these limitations, parents, family members and friends must bear the burden of chauffeuring minors to a number of places. At the other end of the spectrum, many older adults are also unable to drive. According to recent statistics, older adults generally outlive their ability to drive by nearly a decade on average.[34] With our aging society, this has significant importance as well.

In terms of age and driving, some important trends have been occurring. Have you noticed many teens today could care less about getting a driver's license. Today, more teens are waiting until they are 20 years of age or older before taking their driver's test for a handful of reasons. Many have no car, or means to afford a car, while others simply see no need to have one because they travel by other means. This trend has been further supported by a general lack of funding for drivers' education courses in public schools.[35]

Clearly, being too young or too old limits our ability to get around. But age is not the only limiting factor when it comes to transportation in the country. Millions of people suffer various disabilities that prevent them from driving a car. Did you know one in eight people have some type of disability?[36] These disabilities may be related to poor vision, poor hearing, an inability to walk, or limits in memory and thinking, and a sizable number of these individuals are unable to participate in driving. Instead, like our children and grandchildren,

---

[34] AAA Senior Driving. "Facts and research." Website, 2017. Retrieved from http://seniordriving.aaa.com/resources-family-friends/conversations-about-driving/facts-research/

[35] Bush, Evan. Teens delay getting licenses, and their driving is worse. *Seattle Times*, 2016. Retrieved from http://www.seattletimes.com/seattle-news/transportation/young-drivers-wait-to-get-licenses-with-dangerous-consequences/

[36] Cornell University. "Disability statistics." Website, 2015. Retrieved from https://www.disabilitystatistics.org/reports/acs.cfm?statistic=1

they rely on public transit in some cases and rides from family members and friends in others. Here again, our transportation system fails to meet our needs.

All told, about a third of the entire American population does not drive a car. Out of every 1,000 residents in the U.S., fewer than 700 drive. [37] Whether you're too young or old, too disabled, or simply have too few resources to afford to drive, you belong to a sizable segment of the population that receives little benefit from our current mode of travel. Even if you throw Uber, Lyft and other "ride-hailing" programs into the mix of transportation options, millions are still underserved because of user and accessibility issues. Try getting into a standard car from a wheelchair or while using a cane! Our transportation system is hardly fair to these individuals. Despite the fact our nation prides itself on equal opportunity, opportunities for transportation in the U.S. are anything but equal.

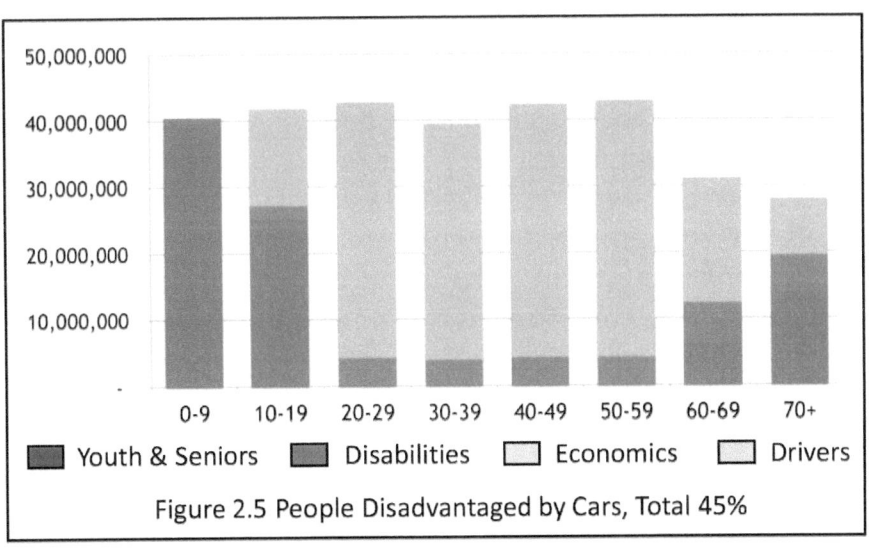

Figure 2.5 People Disadvantaged by Cars, Total 45%

---

37 U.S. Department of transportation, Office of Highway Policy Information. "Our nation's highways: 2011." DOT Website, 2014 Retrieved from https://www.fhwa.dot.gov/policyinformation/pubs/hf/pl11028/chapter4.cfm

Our Transportation System and the LAW (Land, Air and Water)
The network of roads and highways in America is an amazing accomplishment resulting from extensive planning, dedication and hard work. Major architectural and engineering feats have allowed us to drive our cars places that were once considered off limits. Just think about the "Big Dig" in Boston or the Golden Gate Bridge in San Francisco. But at the same time, this network of roads takes up a pretty sizable area. In fact, America's paved areas occupy more than 39 million acres. That's more than the entire state of Georgia! If you were to put a price tag on this land, it would be nearly $500 billion dollars based on conservative estimates.[38] Of course, this fails to consider the fact that many roads and highways are in major cities where land is at a premium. When you think about roads in this way, things don't seem quite as peachy.

Cars and trucks demand roads on which to drive. All forms of transportation, in one way or another, require some type of infrastructure. But what if the infrastructure was much less extensive? What if it was more environmentally friendly? In addition to occupying 39 million acres of land, our roads and highways have several negative effects on the environment. By nature, water cannot be absorbed through road surfaces but are instead diverted away to make travel safer. But the rain and water runoff from concrete and asphalt add sediment, chemicals and debris to the water we potentially drink while contaminating lakes, rivers and streams.[39] In addition to requiring a significant area of land, paved roads affect a much larger area when these things are considered.

---

[38] Larson, William. "New estimates of value of land of the United States." Bureau of economic analysis, 2015. Retrieved from https://www.bea.gov/papers/pdf/new-estimates-of-value-of-land-of-the-united-states-larson.pdf

[39] Nemeth, Andrew F. "The effect of asphalt pavement on storm water contamination." Worcester Polytechnic Institute, 2010. Retrieved from https://web.wpi.edu/Pubs/E-project/Available/E-project-052810-151011/unrestricted/Asphalt_and_Stormwater_IQP_2010.pdf

Some of the same chemicals and pollutants that end up on our roads and in our drinking water can also end up in the air we breathe. You likely know about the harmful effects that lead, nitrogen oxides, and carbon monoxide have on the environment as well as on our own health. But did you know cars and trucks account for three-quarters of all carbon monoxide pollution in the U.S.? In fact, according to the Environmental Protection Agency, half of all air pollutants come directly from the vehicles we drive and ride.[40] Even after decades of efforts to "clean up" fuels and engines, cars are still the major contributor to air pollution.

When thinking about the environmental impacts of our transportation system today, you may wonder about the potential benefit electric cars offer when compared to our traditional cars and trucks. Certainly, electric cars are better for the environment, and who wouldn't want to own a Tesla? While over half a million have been sold in the U.S., these cars still account for less than one percent of the cars on the road.[41] Likewise, they still require an abundance of paved roads in addition to new infrastructures like charging stations. Electric cars have been a step in a good direction, but a more dramatic change is needed if real environmental benefits are to be realized.

As you can see from the information discussed in this chapter, our current transportation system has some major issues. Unfortunately, we may not be consciously aware of these problems on a daily basis. Many of us wake up every morning, get in our cars, take the kids to school, and then head off to work. We may sit in traffic for a while, make a stop to get fuel, and in all likelihood, check a few texts and emails along the way. And when comparing

---

40 U.S. Environmental Protection Agency. "The plain English guide to the Clean Air Act." Website, 2015. Retrieved from https://www.epa.gov/sites/production/files/2015-08/documents/peg.pdf

41 Cobb, Jeff. "Top six plug-in adopting countries – 2015." *Hybridcars.com*, 2016. Retrieved from http://www.hybridcars.com/top-six-plug-in-vehicle-adopting-countries-2015/

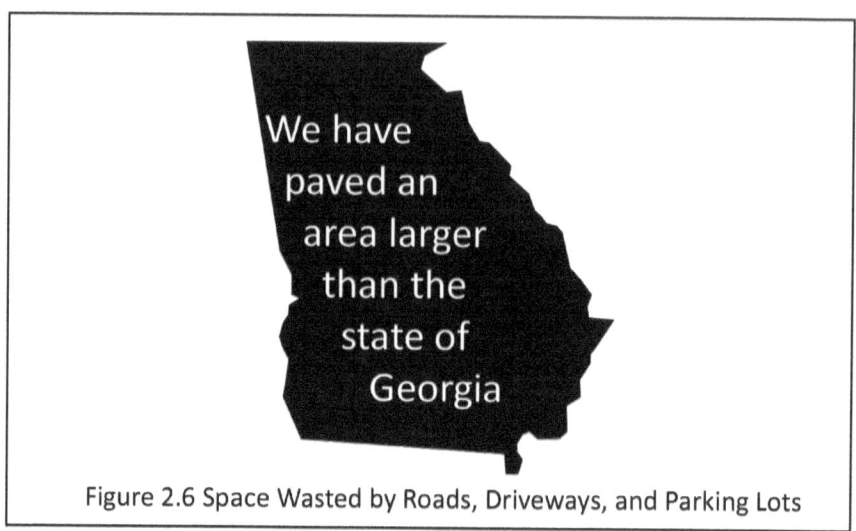
Figure 2.6 Space Wasted by Roads, Driveways, and Parking Lots

our travel today to decades past, things may seem pretty good. But in many ways, our transportation system fails in meeting our needs on a daily basis.

In addition to the financial costs experienced in transportation today, we pay dearly in time and opportunities when we travel from one destination to another. We also are exposed to significant threats to our health and safety. And the impact that our transportation system has on the environment and our energy resources is tremendous. All the while, nearly half of the population is underserved in terms of their transportation needs.

While these issues sound rather gloomy, there is a silver lining. Today's technologies and advances offer creative solutions to these problems. From new types of vehicles to innovative infrastructure design, many of the shortcomings of today's transportation can be alleviated if not eliminated. That's the good news! Exciting alternatives exist, and they have the potential to not only improve our ability to get from place to place but to also enhance our quality of life in so many other ways. Hang onto your hat…it's about to get fun.

3
Designing the Ideal Transportation System

DURING THE HEIGHT of the Cold War, an interesting chain of events occurred that involved the lighting industry in the Soviet Union. At that time, the Soviet Union established five-year plans for a number of sectors, and the lighting industry had its own set of measures that defined its success. Specifically, light fixture production goals were determined by the total weight of the fixtures made during a given time. However, the Soviet lighting fixture factories were unable to meet their assigned quota if they made the traditional mix of light fixtures.

Facing this dilemma, and likely some political ramifications as well, the light manufacturers came up with a grand solution...simply make each individual light fixture heavier. By making heavier chandeliers, they achieved total weight quotas while greatly reducing the number of light fixtures made. It seemed like a win-win for everyone. Unfortunately, they soon realized there were not enough total light fixtures to go around, and many ceilings collapsed, unable to support the weight of the heavy chandeliers. What seemed like a good idea turned into a fiasco leaving many people literally in the dark.

For any industry, effective design is dependent on the right set of goals. The weight quotas for the Soviet lighting manufacturers seemed reasonable at the time, but solutions focused on those goals led to a less than ideal result. So, what goals should we have for the ideal transportation system? If we are going to "get it right," we need to consider specific goals that would create the perfect system to get us from place to place. Having identified many of the problems with our existing transportation system in the last chapter, we can start to think how these can be eliminated through better design. But before we take this leap, establishing key targets and goals that define a more perfect system is important.

In this chapter, we explore a number of aspects that define the perfect transportation system. From vehicles to environmental effects, a variety of features can be considered in determining what defines the best possible system. And these features can then be used to create specific goals within the industry. Like the Soviet Union's lighting industry during the Cold War, our transportation system has strayed away from the ideal, and it's time for us to take a fresh look at these issues to get us back on the right track.

Transportation's Obesity Epidemic

You are likely aware that America as well as many other countries are facing an obesity problem. But did you know our transportation industry has suffered a similar issue for decades? If you recall from the last chapter, a single tablespoon of oil can carry a pound of oranges all the way across the country by freight train, but the same tablespoon will not even allow our cars to get out of the parking lot. Why? In part, this reflects poor energy efficiency of our current automobile engines, but the bigger problem is the weight of our cars. In other words, our transportation system needs to go on a diet!

This problem mostly has to do with simple physics. When we travel from one place to another in our cars, energy is required to overcome various types of resistance. Think about bicycling. If you have ever biked on a windy day, you quickly appreciate a tailwind and dread a headwind. Likewise, you have no problem understanding cycling up a hill or mountainside requires a significant increase in effort. And when you want to accelerate you have to push harder. Cars are not any different. Air, road, incline, and acceleration resistance all affect the amount of energy required to get us to our destination. And while the weight of the car does not affect air resistance tremendously, weight is directly related to the resistance encountered from surface and steepness of the road and accelerating up to speed.

Most cars today weigh about two tons...or 4,000 pounds. That's a lot of weight! Over time, the weight damages the roads on which we travel, and increases the amount of damage suffered during a crash. In fact, a 10 percent increase in vehicle weight is associated with a 40 percent increase of the wear and tear on road surfaces.[42] Weight also affects other components of the car. Heavier weight demands more complicated (and heavier) braking and suspension systems, larger engines, as well as more advanced safety systems. And by increasing fuel consumption, heavier weights also cause higher amounts of air pollution. However, if we reduce the weight of one component, then other components can be lighter, spiraling to create a virtuous cycle of improvement in weight and other effects.

When we start to think about the ideal vehicle for transportation, a key issue thus concerns the vehicle's weight. In general, cars weigh about 20 to 40 times as much as a single passenger. But what if our vehicles' weight was dramatically less? Resistance would decline resulting in reduced energy demands, and a number of other weight burdens that now exist on our cars would no longer be needed. Not only should transportation vehicles be safe, energy efficient,

---

[42] Karim, Mohamed Rehan, Nik Ibtishamiah Ibrahim, Ahmad Abdullah Saifizul, and Hideo Yamanaka. "Effectiveness of vehicle weight enforcement in a developing country using weigh-in-motion sorting system considering vehicle by-pass and enforcement capability." *IATSS research* 37, no. 2 (2014): 124-129.

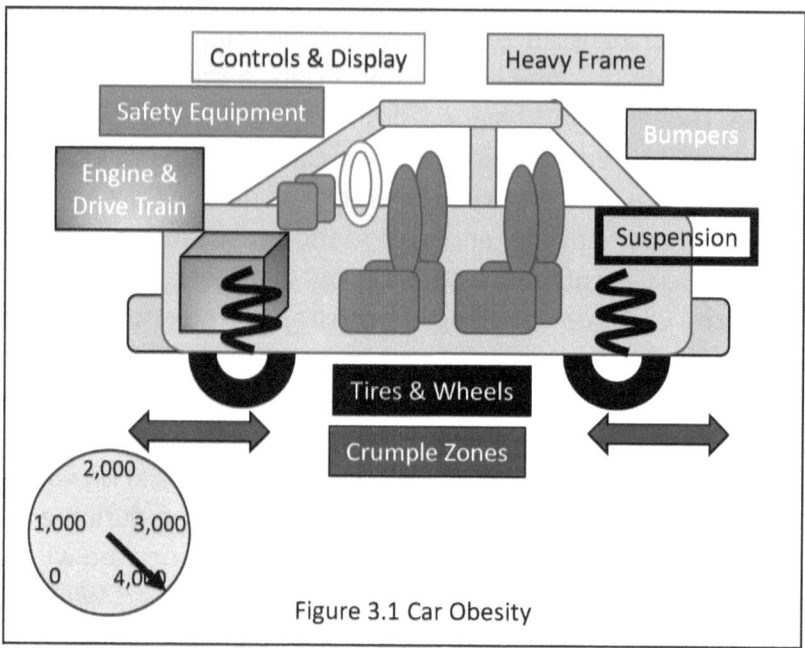

Figure 3.1 Car Obesity

comfortable and have the needed space to carry us and other things to our destinations, but they should be lightweight. In fact, the ideal car should weigh less than the load it is supposed to carry.

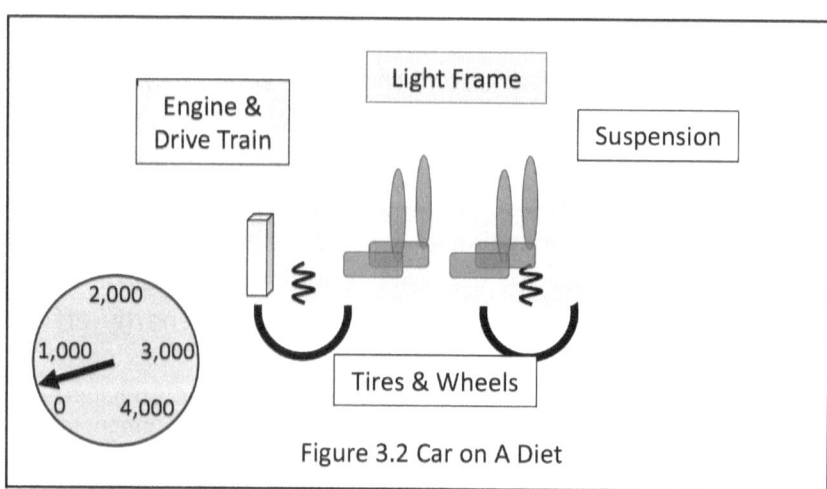

Figure 3.2 Car on A Diet

You may think reducing your car's weight to this degree might be impossible. After all, the car's weight would need to be reduced by 95 percent in order to achieve this goal. But let's think about this concept from a different perspective. If we started with a bicycle, we could increase the bike's weight by 400 percent and still achieve our objective. You may have noticed a number of bicycles that have added an electric motor to their design to allow easier transportation up hills or to allow the cyclist to rest when tired. Even with these electric motors and batteries, the weight of the bicycle is still much less than its passenger, and it allows healthy exercise as well. From this viewpoint, the possibility of designing a vehicle with our targeted weight goal is possible.

Today's cars are certainly comfortable with many bells and whistles, but in many ways, they remind us of the old Soviet chandeliers. Design has gone well past the practical needs of transportation to include many luxuries and amenities we have come to expect. We seriously think a driver's seat is more comfortable than a couch at home! But this type of design is far from ideal. The costs the additional weight imposes on transportation effectiveness and efficiency are tremendous. And like the ceilings from which the Soviet lights hung, our transportation infrastructure pays a high price in the process. Therefore, one primary goal for a new transportation design would be to create lighter weight vehicles that facilitate improvements in other transportation system areas.

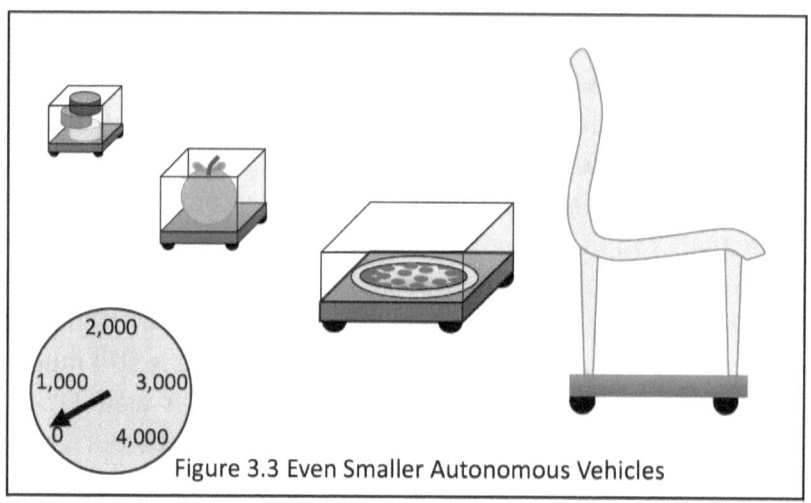

Figure 3.3 Even Smaller Autonomous Vehicles

The Dimensions of Time and Space

Whether you're a Millennial, GenX, or from our generation, some things transcend time and space. Take *Star Trek* for example. Gene Roddenberry's conceptual masterpiece has spanned more than 50 years and been responsible for over a dozen films and hundreds of television shows during this time period. In respect to travel, Star Trek also highlights one of our most significant desires for transportation…immediacy. How cool would it be to say, "Scottie, beam me up," and in an instant, be at your destination? Though clearly still science fiction in our world today, teleportation would finally conquer that dreaded nemesis so many of us experience… traffic.

When designing an ideal transportation system, time efficiency is critically important. The average American loses almost half an hour getting to work each day, and this is even worse for those with longer commutes or for people living in congested cities. The ideal system would allow us to not only get to our destination as quickly as possible, but it would also allow us to have free time along the way, avoid unnecessary stops, and allow easy access. Each of these areas has important implications for new transportation system concepts.

Let's consider the layout of our current transportation system first. For many large cities, transportation is on a grid-like pattern. In other words, roads and transit systems move along corridors that travel either north-to-south or east-to-west forming a kind of checkerboard of transportation. Other cities use a hub-and-spoke pattern that looks like a spider web. In other words, the hub lies at the center of the web, and everything is heavily dependent on hub activity for efficiency. And of course, some cities have a combination of these two patterns of transportation, such as Washington, DC.

Unfortunately, neither grid nor hub-and-spoke patterns are very efficient. Both systems have a higher concentration of travelers in city centers causing delays, congestion and unnecessary stops for many individuals. As a result, these layouts often have bottlenecks located in central "hub" areas resulting in delays and congestion throughout the entire transportation system. If you have ever waited on an outbound subway traveling from the heart of New York City at five o'clock rush hour only to see car after car being too crowded to ride, you appreciate what we are talking about. And should one of these lines break down, then you can certainly expect a rough evening ahead.

While the example of the subway system in New York depicts these problems well, the same problems related to traffic and congestion also result on the roadways with grid and hub-and-spoke patterns. Access to on and off ramps for highways close to city centers is a common problem during peak travel times. As you inch along trying to get onto the Interstate on your way home, precious time is being wasted. Knowing this, we try to steal brief seconds of time on our smartphones, but this is hardly high-quality time. Is it any wonder why we are all becoming the victims of Attention Deficit Hyperactivity Disorder (ADHD)?

Unfortunately, these transportation layouts are not simply inefficient because of their design...they also have negative effects on the way cities grow. It's essentially a Catch-22. Transportation patterns evolved because of a concentration of people in the center of urban areas. But by meeting these needs, the patterns themselves encouraged further concentrations of people in the same places. Why is the property value of most downtown cities more expensive? In part, this is because jobs, restaurants, and stores migrate to these areas. But at the same time, avoiding long commutes, traffic, and transportation bottlenecks further encourages these trends. All the while, urban areas become increasingly congested, more expensive in which to live, and more challenging regarding transportation solutions.

With this in mind, an ideal transportation system must better address space planning, especially in highly concentrated areas. In order to do this, we have to think about the places where people want to travel. If you think about what we do with our time, we tend to move in one of two directions. Certainly, jobs, restaurants and stores represent the places we seek often for obvious reasons. But what about the other times? On weekends, vacations and even some evenings after work, many people seek out remote places. Beaches, mountain trails, lakes and other natural environments have an inherent allure for most of us. In this regard, ideal transportation systems must accommodate this dichotomous nature of our behaviors.

This idea to improved space planning and transportation is not new. At the Savannah College of Art and Design (SCAD), a group of faculty, students and alumni spent 10 months developing urban dwellings whose footprint occupied a single parking space measuring 16 by 8 feet. Noting that the U.S. has 5 parking spaces for every car in the country, an incentive was identified to better utilize these spaces. Interestingly, however, their designs encompassed a common green space for the dwellings (called SCAD pads) to address our need for nature as well as urban environments.[43]

While SCAD's focus was on improved use of urban spaces, it highlights the importance for transportation systems to not only

---

[43] Savannah College of Art and Design (SCAD). "What is SCADpad?". Website, 2014. Retrieved from http://www.scadpad.com/what-is-scadpad

ensure transit to urban areas but to natural environments as well. It also highlights the excess space our current transportation system requires. Even Apple's iconic new office complex, which will be described shortly, has devoted more space to parking than it has to actual usable space. When thinking about goals for the ideal transportation system, we need to critically evaluate the demands various features place on the system as a whole. This is particularly important when it comes to space and time.

When it comes to scarce resources, time is one of the most critical ones we have today. And as it pertains to urban areas, space is equally important. But what if new transportation designs permitted us to have more of both of these scarcities? Instead of spending hours a week in your car concentrating on the road, what if you had that time to do other things? Rather than commuting long distances to escape the city for the peace and beauty of nature, what if designs allowed both to be experienced in close proximity to one another? When it comes to setting goals for the ideal transportation system, these considerations regarding time and space are essential.

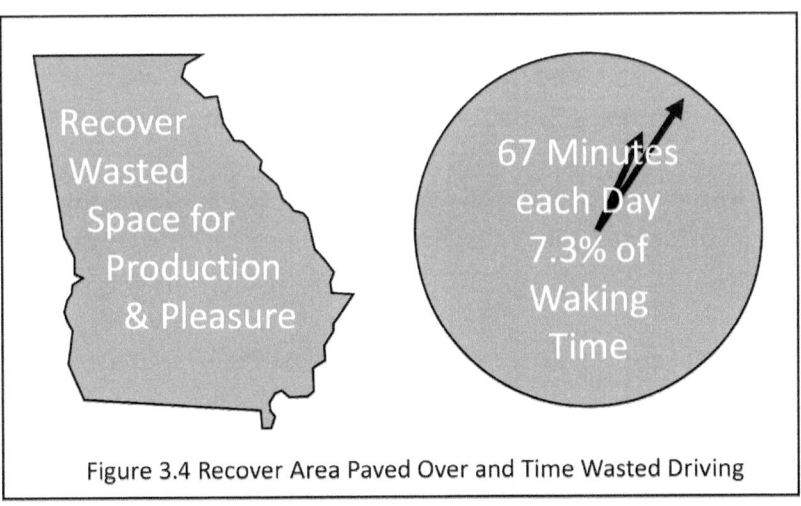

Figure 3.4 Recover Area Paved Over and Time Wasted Driving

First Do No Harm

On our first trip to Japan many years ago, we were invited to the home of our host. During the evening, our host drank freely and was a bit intoxicated by the time we needed to return to our hotel in Tokyo a few hours later. As a result, his wife was nice enough to drive us there while her husband slept in the back seat. It was during our trip back to Tokyo that we learned the penalty for the first offense of intoxicated driving was up to five years in prison in Japan![44] While the laws against driving under the influence are pretty tough in the U.S., they are by no means as strict as Japan's. But given the fact that alcohol is involved in roughly a third of all car crashes involving a fatality, such punishments can be appreciated.[45]

Safety is a primary concern for any industry, and the transportation sector is no different. After years of decline, in 2016 deaths in car crashes in the U.S. rose to more than 40,000, and over 4 million people were injured.[46] Therefore, personal safety is a major area of concern when developing the ideal transportation system. Because human error is involved in more than 90 percent of car crashes, we have one of two options to improve personal safety on a large scale. We can either construct devices and accessories to prevent or counteract human error, or we can eliminate the human component altogether. So far, the trend has been the former.

Like Japan, the U.S. has adopted laws seeking to reduce human error, such as those associated with drinking and driving, texting and driving, and even age-related impairments. Other strategies involve gizmos to offset our mistakes. Rearview cameras, lane sensors, other vehicle sensors, and airbags are just a few of the common accessories

---

44 Svan, Jennifer H. "Japan's tougher drinking, driving laws take effect." Stars and Stripes, 2007. Retrieved from https://www.stripes.com/news/japan-s-tougher-drinking-driving-laws-take-effect-1.68979#.WRCuTuXyvIU

45 Centers for Disease Control (CDC). "Impaired driving: Get the facts." CDC Website, 2017. Retrieved from https://www.cdc.gov/motorvehiclesafety/impaired_driving/impaired-drv_factsheet.html

46 National Safety Council. "NSC motor vehicle fatality estimates." *NSC Website*, 2017. Retrieved from http://www.nsc.org/NewsDocuments/2017/12-month-estimates.pdf

on many cars to enhance personal safety. But despite these efforts, human error still causes a significant number of deaths and injuries on the road year after year. In part, driverless cars have recently attracted increasing amounts of attention and interest because of this.

So what accounts for car crashes that are not related to human error? A small percentage of car crashes are due to problems related to the car itself. Other causes extend beyond the car and the driver to the surrounding environment. If you live in northern climates, you may have experienced a "white-out" while driving, or you may have suddenly found yourself in a downpour or fog not being able to see a few feet in front of you or skidding on snow and ice. Weather-related conditions can certainly undermine personal safety not to mention renegade deer, fallen tree branches, and the ever-present pothole that might cause a blowout or loss of control of your car. We've even seen the occasional extension ladder lying in our path (though not listed as a common road hazard!).

Safety concerns involving a transportation system pertain to more than the vehicle and driver. These concerns also relate to road surfaces, road debris, road designs, traffic signals and a variety of other areas. While these may make up a minority of the causes for car crashes, they are important nonetheless. Therefore, environmental hazards must be taken into account when designing an optimal transportation system. In considering an ideal system, designs of infrastructure that complement vehicle transportation must also be contemplated. In some cases, poorly designed infrastructure can affect our safety as much as a poorly designed vehicle.

This brings us to safety of a different kind. Our immediate personal safety is certainly important, but environmental safety is also critical because it affects us in the longer term. Clearly, our current transportation system is lacking in this regard. In addition to the damage that fuel emissions and greenhouse gases cause to air quality and climate change, roadways and cars also contribute significantly to pollution and contamination of our lakes and rivers. And the accumulated waste byproducts of old cars and tires lead to additional areas of concern. Though many old car parts are sold as scrap and recycled, thousands of junkyards across the country contribute to land waste, water waste and air pollutions as well.

When we start talking about designing a new transportation system, the goal should be to first do no harm. This means avoiding harm to us as individuals and to others on the road, and it should also address preventing harm to us through our environment. The ideal transportation system should allow us to travel from one place to another while reducing the chance of human error and disruption of our surroundings. Likewise, potential threats along the way should be minimized if not eliminated as well. In these areas of design, we have a long way to go from where we are currently.

Naturally Fitting In

For a few years now, Apple has been building its new 175-acre facility in Cupertino, California. Without question, the futuristic, spaceship-appearing structure looks to be well ahead of its time even at first glance, but its innovations go well beyond what meets the eye. Despite its massive size, Apple Park will operate using 100 percent renewable energy. In addition to a 17-megawatt rooftop solar system, a natural ventilation system harnessing the benefits of natural wind will eliminate the need for heating and air conditioning nine months out of the year! Likewise, it will utilize recycled water while sharing this resource with the local community of Cupertino.[47] Certainly, the energy demands of a company like Apple are enormous, but by using existing, natural resources readily available in its immediate environment, Apple Park will achieve remarkable energy efficiency and sustainability through its innovative and logical design. Apple established ideal goals from the start and is setting out to achieve them.

---

47 Anderson, Casey. "New Apple headquarters sets records in solar and green building." Renewable Energy World, 2017. Retrieved from http://www.renewableenergyworld.com/ugc/articles/2017/02/28/new-apple-headquarters-sets-records-in-solar-and-green-building.html

This is not the case with America's transportation system. First of all, the energy used to fuel our cars is hardly readily available. The transportation system uses over 14 million barrels of oil every day![48] Cars, light trucks and delivery trucks use roughly 85 percent of this fuel, yet these same vehicles operate at an energy efficiency of 20 percent or less.[49] Energy losses mostly occur due to the inefficient nature of automobile engines, which account for three-quarters of the total energy lost. But additional inefficiencies exist due to the mechanical operation of the car, wind resistance, and friction.[50] Our cars are therefore double-edged swords using fuel we don't readily have while wasting its use through normal transportation activities.

So why hasn't America adopted more energy efficient means of transportation? If Apple can have a mega-office that runs completely on renewable energy, why are we still using fossil fuels for our daily commutes? Good question. Certainly, alternative options have been available. For example, the electric car is much more energy efficient and environmentally friendly, and in its modern form, it has been around since the 1990s. However, as highlighted in the documentary *Who Killed the Electric Car?*, oil companies, car manufacturers, car part suppliers, and politicians with various biases undermined the adoption and potential success of this more efficient vehicle.[51]

In contrast to our gasoline-powered cars, electric cars have an energy efficiency that exceeds 80 percent.[52] Having previously leased Chevy Volts and now owning Teslas, we must admit it is a great feeling not to

---

48 American Energy Independence. "American fuels." Website, 2013. Retrieved from http://www.americanenergyindependence.com/fuels.aspx
49 University of Washington. "Improving internal combustion energy efficiency." Website, 2010. Retrieved from http://courses.washington.edu/me341/oct22v2.htm
50 Ibid.
51 Paine, Chris (Director). Deeter, Jesse (Producer). "Who Killed the Electric Car?" Film, 2006. Sony Pictures Home Entertainment.
52 Stricklett, K. L. *Advanced components for electric and hybrid electric vehicles.* US Department of Commerce, National Institute of Standards and Technology, 1994.

fuel our cars with gasoline or oil. But the advantages are so much more. For one, electric cars do not need to idle when stopped, and therefore, precious energy is not being wasted and air pollutants are not being emitted. In addition, the regenerative braking system on electric cars allows some of the energy being used to be reclaimed. Instead of applying brakes and wasting the energy of motion, this kinetic energy is used to recharge the car's battery. These advantages are being recognized by consumers as well. Since 2012, electric car sales have been averaging increases of 50 percent or more globally. And in some countries, electric cars on the road are over 5 percent of all vehicles.[53]

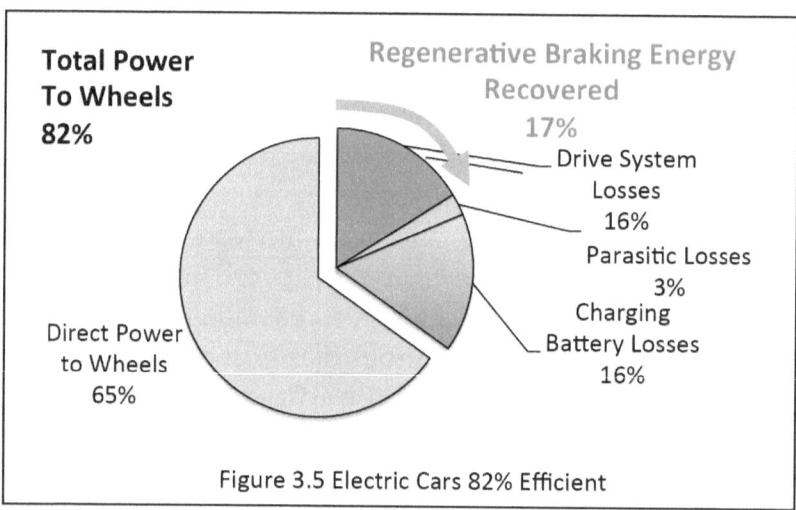

Figure 3.5 Electric Cars 82% Efficient

Electric cars are not perfect either, however. The batteries that are required for these vehicles use lithium, which is a scarce resource itself. In addition, these batteries account for a significant amount of weight in the car. In the Tesla Model S, the lithium battery is roughly

---

53 Romm, Joe. "37% of Norway's new cars are electric. They expect it to be 100% in just 8 years." ThinkProgress.org, 2017. Retrieved from https://thinkprogress.org/norway-aims-to-end-sales-of-fuel-burning-car-by-2025-as-ev-market-soars-edeac854f1e

28 percent of the car's total weight.[54] And electric cars still require roads and highways, which contribute to water pollution and other hazards. Regardless, battery-operated vehicles are a major step forward in energy efficiency and sustainability when compared to traditional automobiles. Just imagine the impact we could have immediately on our energy usage and on the environment if everyone drove an electric vehicle today!

So, what is the takeaway here? In addition to designing a transportation system that minimally impacts our surrounding environment, we should also take advantage of the natural resources of energy that our environment provides. Apple is a leader in this regard as are many other companies, and an ideal transportation system should do the same. Using natural energy sources like wind, solar and electrical power is not only more sustainable and environmentally friendly, but it also is much less costly over time. And by designing vehicles and infrastructures that are energy efficient, we can get the most out of our efforts.

Specific goals include, the vehicle should weigh less than the load, no one should waste time driving a vehicle, the system should be more efficient than a freight train. There should be no fossil fuels, no runoff of polluted water, and no pollution from vehicle exhausts. Deaths and injuries from transportation should be eliminated. People and packages should be delivered promptly, efficiently at low costs. Travel should be fun.

To design new and improved transportation systems, an ideal standard needs to be identified. In this chapter, we have identified many of the key components of a transportation system that can define such standards and raise the bar when it comes to design. Transportation vehicle weight and energy efficiency are certainly important facets as are issues related to personal and environmental safety. Likewise,

---

54 Teslarati Network. "Tesla Model S weight distribution." *Website*, 2013. Retrieved from http://www.teslarati.com/tesla-model-s-weight/

environmental impacts and space considerations should be included as well as the ability to minimize time investments for traveling. While achieving perfection in all of these measures simultaneously may be difficult if not impossible, striving for perfection is what fuels innovation, creativity and monumental change.

Figure 3.6 Goal: Universal Mobility
– Everyone, Everything, Everywhere, Fast, Fun, Frugal and Safe

From this perspective, we can now offer some revolutionary and dramatic possibilities in a new transportation system design. From new concepts in vehicle styles and passenger travel to radical changes in transportation infrastructures, we can greatly improve today's transportation system. At the same time, we can come much closer to achieving the ideal standards discussed in this chapter. Equipped with the major advances in technology and science today, these goals are not simply concepts but realistic possibilities for the immediate future. And, as you will soon see, envisioning these changes is both satisfying as well as exciting.

4
Autonomous Vehicles

WARREN: MY FATHER loved to walk. He would regularly take long walks every day. But after my stepmother contracted Parkinson's disease and could no longer walk with my father, he didn't walk nearly as often because he didn't enjoy walking alone. My stepmother would have considered a motorized scooter, but she lacked the fine motor skills for this to be feasible. Good options to help her get around were quite limited, and as a result of her immobility, my father's level of activity declined too.

Seeing my father and stepmother experience the limitations of our current transportation system prompted me to start thinking about other options for getting around. How great would it be if everyone had access to a personalized vehicle that took them wherever they wanted? You're probably saying to yourself, "We already have that…it is called a car!" But as discussed, large segments of the population cannot drive a car, and many more have difficulty even riding in one. Automobiles also fail to address mobility needs within buildings and other limited spaces.

My initial thought was to have a personalized mobility vehicle for everyone that provided transportation regardless of their technical abilities and skills. Those with disabilities, children, and older adults with impairments would all benefit from such a vehicle. Of course, you are still encouraged to walk, run, bike and exercise … but now you can have the company of someone who cannot match your gait. Naturally, this remains an important focus for any new transportation system.

What if such a vehicle was also self-driving, that is autonomous? And what if these same types of vehicles could be adapted and used for the transport of a variety of goods as well as people? For example, nearly everyone could benefit from a number of deliveries such as mail, meals and medications. These vehicles could be used to provide these services as well. In fact, these vehicles could enhance many areas of our lives. And the whole transportation system has to be cheap enough so everyone can afford to use it.

In our quest to take advantage of the 30,400:1 opportunity we explore many avenues, including low overhead weight, packing, reusability, and standards. Toys already show the way: remote control cars for $29, a tiny quadcopter for $49, and game controllers that sense your every motion. Tesla has announced a semi with the capacity to pull the weight of 100 people, and we will see how the $150,000 cost includes many things not needed.

With this in mind, this chapter seeks to address the basics of the transportation vehicles that can best meet our needs in today's world. To some extent, history can provide us with some valuable lessons in transport designs, but at the same time, the models we are proposing use today's technologies. From the transport of goods to people, new approaches can be considered. This includes the benefits of having autonomous control of these vehicles. In the process, we can address many of the problems associated with today's transportation system vehicles.

Designing Transportation Around Our Needs

When many people hear the word transportation, they immediately think about cars, buses and subways. Certainly, these vehicles meet important transportation needs. They allow us to get to work, the store, the theatre, and many other desirable locations. Others think more broadly about transportation in terms of supplies and products.

Transportation of goods by way of trucks, trains and freight liners are similarly important. But we would like to create a vision that is even more encompassing. This vision invites you to consider transportation of anything and everything from one place to another. From pills to people and from pizzas to patio furniture, an effective transportation system should address all of these needs.

Warren: My father resided within Charlestown, a retirement community that offered many exceptional services. One of these services offered residents the opportunity to have food delivered to their doorstep from the community restaurants for a nominal surcharge. As my stepmother became less able to walk, going out to a restaurant became increasingly difficult. But she still enjoyed many of the restaurant's foods, and because of this, my father and she often took advantage of this food delivery service. Unfortunately, the food usually arrived late and rather cold making the experience less than they had envisioned.

So, what are our transportation needs? Certainly, Warren's stepmother would have liked to have had more effective transportation to get her to the restaurant in the first place. However, she would have also liked to have received meals at home on occasion at just the right moment and at the right temperature. While this may happen with today's transportation systems on occasion, most of the time, we are waiting on the delivery, and the meal's presentation is less than ideal. The basic need of having food arrive at your home is achieved, but you can appreciate that there is significant room for improvement.

Warren: A few years ago, I purchased a new refrigerator. While determining which model was the right one for my family was a bit of a process itself, getting the new refrigerator to the kitchen was another. A delivery time had to be arranged (of course, not an exact time but a four-hour window), and installers had to be dispatched to our home. At the same time, they had to disconnect the old refrigerator and haul it away. Connecting the water supply was a separate job for a plumber. Again, the basic need of transporting a new refrigerator to our house was achieved as was taking the old one away. But what if the transportation was automated instead of requiring drivers? What if an exact time was possible for the delivery? And what if the installers could also arrive at the exact moment of the delivery through an automated transportation system as well? This again shows how many of our transportation needs are not currently being met.

The transportation system we are proposing goes well beyond personal travel and delivery of your online purchases. It also goes well beyond the delivery of freight and supplies, although these are certainly an important feature. Our proposed system addresses these transportation needs and greatly improves them in the process by making them more efficient, timelier, and less costly in terms of human resources. And it extends this model to the transportation of essentially everything that needs to get from one place to another. Whether it's jewelry, books, mail, groceries or even refrigerators, a transportation system should be able to get these items and ourselves to the exact destination we want at just the right time. These are the real needs of a transportation system.

In considering all of these potential applications described, we propose a model that is dynamic and flexible enough to transport something as small as a prescription dose to something as large as a modular home. Similarly, it can accommodate the specific needs of each of these items whether they might need refrigeration, a sterile environment, enhanced security or human companionship. And while some degree of standardization is required in our model, it also allows for great variety in services, designs and aesthetics. All the while, it greatly improves efficiency, usability, and quality when compared to our current transportation structures.

As a starting point, we first addressed how an automated transportation vehicle could serve all of these transportation applications and needs. Clearly, the same vehicle that carried a prescription dose would not be the same as one carrying a hot meal from a restaurant. Likewise, one designed to transport mail and books could not be used to deliver a refrigerator (and it would be rather silly to send mail and books in a refrigerator-sized vehicle). Because of this, the autonomous vehicles in our proposed system comprise a family of vehicles, each designed for specific contents and specific needs. With this in mind, let's consider the structure that will actually hold the items or persons being transported.

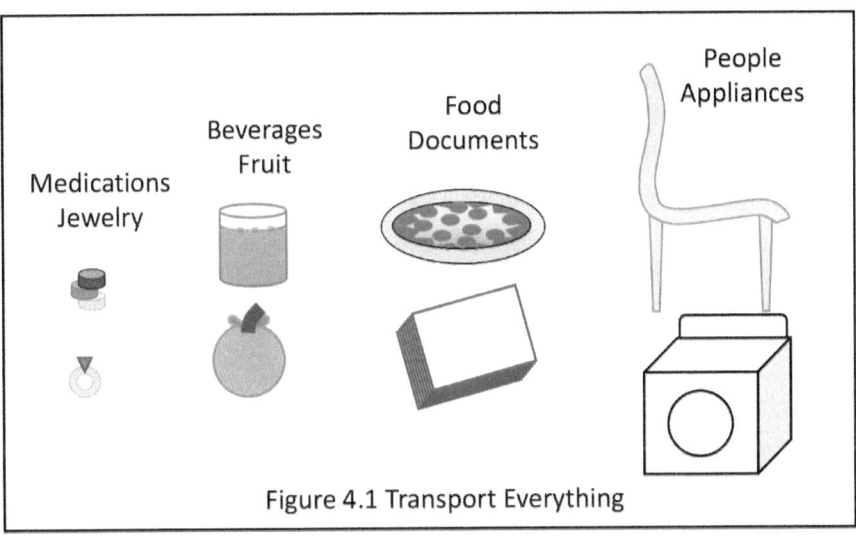

Figure 4.1 Transport Everything

Let's Talk About Containers

You might not think a conversation about containers isn't too enlightening or even interesting. But we expect at least a few cargo handlers over the years wished such a conversation had taken place so that their lives could have been significantly improved. For many centuries, a number of civilizations used various types of cargo containers to trade goods. Merchants from countries like Rome, Greece, Spain, Portugal and Britain required shipping containers for the bulk of their trade when shipping goods overseas.[55] The only problem was

---

55 World Shipping Council. "Before container shipping." Website, 2017. Retrieved from http://www.worldshipping.org/about-the-industry/history-of-containerization/before-container-shipping

that containers during this time period were markedly inefficient and cumbersome.

Over time, a number of different containers were built to carry specific merchandise, but many items were still shipped loose, outside any particular crate or box. For example, rope was used to tie together lumber and timber; amphoras were used to transport wine; sacks were used to store coffee beans and grains; and barrels were used to house liquids of various kinds. These containers were often placed into wooden crates or on pallets for loading onto ships. The initial loading process was extremely labor intensive, but this paled in comparison to the amount of effort required to transfer goods from ships to other carriers. Crates and containers were bulky and difficult to maneuver in small spaces, and they had to be broken apart and repackaged when cargo transfers occurred. Not uncommonly, a cargo ship would be docked for loading and unloading for a longer period of time than it was at sea.[56]

You might think this archaic process of loading and unloading wooden crates only applied to the shipping industry hundreds of years ago. But did you know this process remained in place until the middle of the 20th century? As railroads and freight trains became increasingly popular as a form of transportation, the same cargo containers and loading processes were used to transfer goods from ships to trains. Cargo transfers were already slow and time consuming, but additional transfers to trains made these inefficiencies painfully obvious. In addition, loading and unloading accidents, cargo damage, and theft were significant problems. Something needed to change.

Ultimately, we have Malcom P. McLean to thank for a disruptive technology in the shipping industry. Realizing the problems associated with cargo containers and the transfer process between transportation systems, in 1956 McLean designed a cargo container that could move from cargo ships to freight trains to 18-wheelers more seamlessly. Goods could be loaded once into the containers, locked

---

56 Ibid.

and sealed with contents inside, and transported to their destination via different modes of travel without ever being unloaded and reloaded. And because the containers came in uniform large sizes and shapes, space and storage capacities were used more efficiently. Such a simple concept made a tremendous impact on cargo transportation overnight, and it has not been the same since.

Prior to a standardized *intermodal container* (called intermodal because it travels between different transportation modes), many of the goods shipped overseas were unique, rare or exotic. After all, the expense of labor and shipping made it unreasonable for shipping goods of lesser value. But that changed with McLean's container. Walk into Walmart or any other megastore today, and you will see a panoply of goods made in countries throughout the world. Electronics from Japan, T-shirts from the Philippines, and tools from China are some common examples of goods now shipped from afar that used to be made more locally. Not only did these new containers reduce the need for as many dock workers in the shipping industry, but they also had a major impact on globalized commerce overall.

Let's consider some basic economics. Shortly after McLean introduced his new cargo containers, he found it only cost him 16 cents per ton to load a cargo ship due to reduced labor requirements. In contrast, his prior costs of loading loose goods were nearly 6 dollars a ton! We're talking about some dramatic savings! With such a huge reduction in costs, it became feasible to ship goods from overseas. In fact, large manufacturers with inexpensive labor could produce and ship goods to the U.S. more cheaply than local manufacturers could make the same goods. Historical analysis of the effect cargo containers had on globalization shows they boosted bilateral trade arrangements between nations by nearly 800 percent in the first 20 years![57]

---

[57] H., E. "Why have containers boosted trade so much?" The Economist, 2013. Retrieved from http://www.economist.com/blogs/economist-explains/2013/05/economist-explains-14

What does this mean in practical terms? Take a look at bottled water today. When you go to your supermarket, you will see an array of bottled water options. Some are bottled from local springs while others are processed. But also, on the aisle will be imported water… Perrier, Evian, San Pellegrino, and Fiji to name a few. Before the advent of cargo containers, it would have never been so affordable to ship such products to America, particularly for prices that could rival local water bottlers. But with the ability to send many tons of water bottles from foreign manufacturers at a fraction of the cost, these imports were able to compete not only on quality but also on price with locally produced bottled water.

McLean's idea regarding more practical and efficient cargo containers had far-reaching effects. By eliminating labor requirements and multiple steps in the process of transport, the cost of shipping goods fell dramatically. Likewise, the distance goods could be shipped increased significantly since each stop no longer required unloading and repackaging. And, by using reusable containers that could simply be reloaded and used again, shipping expenses were reduced even further. All of these effects supported globalization of goods that changed local marketplaces substantially.

McLean's cargo containers offered dramatic improvements in how materials and goods were shipped, but they had their limitations as well. For one, if the demand for shipping materials in both directions did not exist, the cargo containers often return to their sites of origin empty or pile up at the destination, which is notably inefficient. They also promoted the shipping of disposable goods as well. As shipping costs fell with these new containers, manufacturers throughout the world who could produce cheap, disposable goods now had access to a number of new markets. Using our water bottle example, inexpensively produced plastic bottles could house imported water while lowering shipping weights. Not only did the cargo containers facilitate imports of foreign goods, but they also encouraged disposable containers and packaging to ship those goods with over a billion water bottles discarded every year.

Regardless, these advances in cargo containers were revolutionary. With only a few different sizes of containers used, the transfer of these containers from ship to rail to truck was much more feasible and efficient. At the same time, these same containers could be easily stored outside (rather than inside warehouses) as needed for later distribution of goods. Though McLean's thoughts of a more standardized cargo container seem rather simplistic today, no one in the preceding centuries had implemented such a concept. This shows that progress and change for the better does not always require an idea that is extremely complex or convoluted. Sometimes a simple solution is staring you right in the face.

Autonomous containers will come in a wide variety. They are optimized for the needs of different contents, for example, keeping your beverage cold, or hot, your pizza piping hot and crisp, your document safe and secure, your fragile flowers fresh and shapely. Some people will be happy with just a place to sit, others want privacy, some may want a large display, others need to lie down, and others will travel as a group. The containers will come in standard sizes, for example, 1" cubes for medications and jewelry, 4" cubes for beverages and pieces of fruit, 4" by 12" squares for food and books, 36" square by 60" high for people and appliances, and larger items.

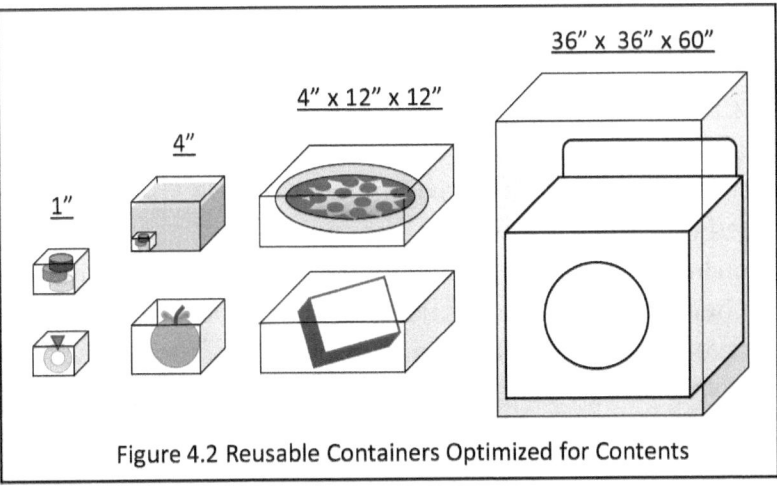

Figure 4.2 Reusable Containers Optimized for Contents

The Concept of Nesting

Have you ever packed the trunk of your car before going on a long trip? If so, you are likely familiar with the concept of nesting. First, you select the essential personal sundries needed for the trip and tightly organize them in a small toiletry kit. The kit is chosen based on these needs as well as the potential space you have in your luggage for the kit. You then pick out clothes, shoes and other items that you will need, but these are also selected based on the amount of space available in your luggage. And then you determine how many pieces of luggage you will take based on the amount of trunk space you have. As you pack your toiletries, your luggage, and other items in your trunk, you are applying the concept of nesting. One item "nests" inside the next, making the most of the space available.

Even before the invention of the modern cargo container, it was well recognized that space came at a premium. Wooden crates and barrels were filled to their capacity in terms of the space they provided. But even so, the actual shipping containers took up a great deal of space. Modern intermodal containers were much more efficient in this regard since their size is much larger and more standardized. This allowed better use of available space on the transportation vessel being used. And through the use of nesting, space within the containers themselves could be better utilized as well. For example, smaller boxes could be placed together in larger boxes, which could then be placed with other similar size boxes into even larger ones before being loaded into the cargo containers. This permitted goods and packages of various sizes to be nested together and shipped to a common destination while maximizing the use of space. This greatly enhanced efficiency of shipping transport.

If you have recently unpacked a package from UPS or Amazon, you can further see how nesting is used as a common strategy. Especially if you have a package containing multiple items, each individual item has its own packaging and is then nested with other packaged items in a larger box. Now consider how these same goods might arrive at a warehouse or distribution center. If we consider our bottled water example again, thousands of boxes of Perrier might arrive in a much larger intermodal container to the warehouse, and within each box, progressively smaller packages of bottled water are nested within. As these smaller boxes are unpacked, they are then shipped to individual stores or customers for final delivery.

As you can see based on this example, nesting not only allows space to be better utilized, but it also allows goods to travel to their destinations at the lowest cost. In essence, some of the bottled Perrier destined to one supermarket has "piggy-backed" with other bottled water packages going to other stores. Likewise, some of these packages might have "piggy-backed" with other items as well. In this regard, nesting is a little like a ride-sharing program, but instead of people carpooling together to get to work, various goods share a ride together in order to reach their site of delivery. In both instances, transportation is made more efficient.

While the use of nesting offers great opportunities for efficiency in the use of space and travel, it does have one nemesis in today's shipping industry…packaging. Think about the last package delivered to your home. It likely arrived in a sizable cardboard box, and when you opened it, you dug through an array of disposable packing materials like Styrofoam peanuts, air-filled plastic bags, or wads of brown paper. Nestled within this packing material was then a much smaller cardboard box that contained your order. And as you opened this box, additional packing materials surrounded your actual item. Of course, your actual item likely had its own plastic packaging that required several minutes of your time to open.

With this example, you can appreciate how some inefficiencies and wastes still exist in the shipping industry even now. Cardboard, paper and plastic packaging has come a long way in the last century, but this type of packaging is hardly reusable or efficient in its use of space. In addition, the use of nesting within containers demands warehouse storage and distribution centers in many instances. Just as loading and unloading goods from wooden crates at dock sites in the past represented another step in the transportation process, warehouses continue to occupy a necessary step in the transit of goods today so goods can be unloaded and re-routed to their final destination.

Despite these limitations, nesting continues to be an effective way to enhance transportation system efficiency. Because of this, it remains an important strategy to be considered in the design of transportation vessels and vehicles. Space continues to come at a premium cost, and nesting allows us to best utilize space when moving something from one place to another. This not only pertains to goods but also people as we will discuss later in the chapter.

Using the example sizes we described above, at least 10 medication containers fit in a beverage container, 10 beverage containers fit in a food container, and 10 food containers fit in a personal mobility container. This gives an overall packing factor of at least 1,000 medication containers to a personal mobility container. Now you can appreciate the power of packing and the enormous potential savings in both efficiency and costs.

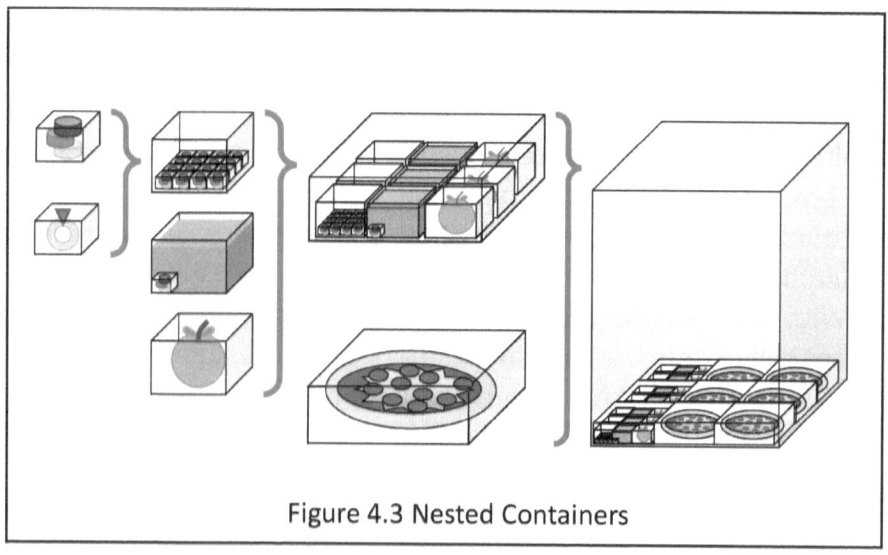

Figure 4.3 Nested Containers

Options of Mobility

Thus far, we have talked about containers and how containers can best be utilized to carry various items through nesting. The next question is how do we get our loaded containers from one place to another. McLean designed intermodal cargo containers based on existing mobility platforms. For example, cargo containers could be easily loaded onto tractor trailer trucks, railway cars, or ocean freighters equally easily. Each mode of transportation had its own family of mobility platforms on which these common cargo containers could be placed and transported to a destination.

In designing a new transportation system, however, we have a unique opportunity to consider not only new containers but also new mobility platforms. The mobility platforms are optimized to make best use of the transportation mode, whether is a flat surface, steel rails, water, an air cushion, magnetic levitation, or flying. Hyperloop would also require working in a vacuum, which is a challenge for both mobility platforms and containers.

Of course, this raises many questions. What size is the container being moved? Does the container need power or communications or air conditioning or other services? How far do we have to move the container? How quickly does it need to get to its destination? What infrastructure is available? The answer to each of these questions affects the type of mobility platform we might consider. And like McLean's cargo containers, having some degree of standardization in mobility platform design offers some potential advantages as well. The interface between containers and mobility platforms is one main area for standards. In addition to the physical size of the container interface, how does the container lock in place on the mobility platform? How are power, communications, air, and other services passed to the container?

Warren: At one point, my father had nearly two dozen different pills he had to take each week, and each medication had a different schedule. It became nearly impossible for him to organize his weekly pill container, not to mention the challenge to then remember to take each pill at the correct time. Of course, my father is not alone. Many people have similar struggles, and as a result, medication doses are commonly missed or taken at the wrong time. Sometimes, this confusion can even lead to very serious problems.

Until recently, the leading cause of unintentional deaths among Americans 25 to 44 years of age was car crashes (in fact, it remains the leading cause of accidental death in ages 5 to 24 years). This has since been replaced by medication overdoses and poisonings in the 25-to-44 age group.[58] Therefore, one practical application for a new container and mobility platform might involve the ability to provide "just-in-time" medication to individuals. In other words, by having an autonomous vehicle deliver medication to someone at the time it was needed, medication overdoses and poisonings could be avoided. That would have made life much simpler (and safer) for my father.

So how might we envision getting these medications to someone at just the right time and place? Assuming the medications were being dispensed from a pharmacy, the first step would be filling a reusable container with the "just-in-time" doses needed. While this could be performed by a pharmacist or pharmacy technician, the number of medications each day that would have to be placed in containers would make this impractical. In fact, many large pharmacies already use automated machines to fill prescriptions. All we need to do is to utilize such an automated process to fill medication needs in a more "real-time" manner.

Now that we have our medications packed within a small container by our automated pharmacy machine, the container must now be sent to the person's home or wherever they are at the appropriate time. With this in mind, a small container with its own small mobility platform is very feasible. In addition, a small, automated vehicle would be more economical than specialized equipment that only carries medications. Just as Kiva robots collect and transport shelves

---

[58] Centers for Disease Control (CDC). "Ten leading causes of death and injury." Website, 2015. Retrieved from https://www.cdc.gov/injury/wisqars/LeadingCauses.html

of items in Amazon warehouses, our medication containers on their small mobility platforms could travel from the medication dispensing area to the shipping area for transport outside the pharmacy. In essence, these small autonomous vehicles would move as part of this entire process without interfering with other pharmacy operations.

You might think autonomous vehicles this small are not economical, or maybe not even feasible. Just search a toy catalog, and you will see many examples of vehicles that already have most of the capabilities our medication vehicle needs. For example, a quadcopter that is 1.6" square for $49, and on sale for $29. With remote control, and sophisticated stabilization, this tiny vehicle can even lift our medication doses. In future chapters we describe more details of the capabilities our autonomous vehicles need, and how toys are already demonstrating solutions, but you can already see that feasibility and cost are not going to be issues.

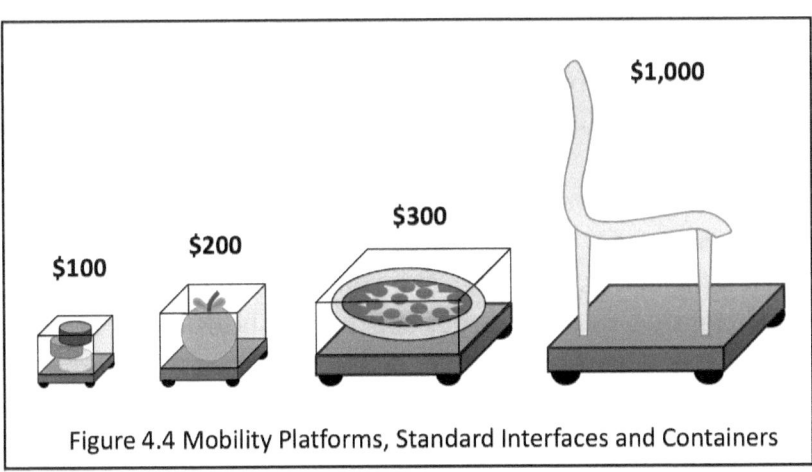

Figure 4.4 Mobility Platforms, Standard Interfaces and Containers

So now we have our medications ready to leave the pharmacy, packaged in a container, and on a small mobility platform. Now what? Let's suppose the eventual destination is across town. It would make little sense for our just-in-time medication delivery to travel all that way by itself. Instead, the small medication vehicle could "hitch a ride" in a larger sized container that had a faster mobility platform optimized to run outside the factory. Using the same concept of nesting that we described for our Perrier water bottles, these smaller medication vehicles could "nest" within larger containers so that space was better utilized. And the larger vehicle would be more efficient in getting our medication to the right place at the right time.

The use of larger vehicles to travel to destinations farther away offers practical solutions as well. With a larger size, these vehicles could travel faster and farther, and the infrastructure available for such travel would be more efficiently used outside of the pharmacy setting. What if the pharmacy was located within a hospital? Perhaps our medication vehicle nests within a moderate sized vehicle with other small autonomous vehicles to get to the perimeter of the healthcare complex more quickly. Once there, the moderate size vehicle then enters an even larger vehicle to speed across town. Once the larger vehicle gets to the neighborhood where the medication is to be delivered, the moderate size vehicle exits the larger vehicle. And as the moderate sized vehicle travels close to the person's home, our small medication vehicle exits for medication delivery…just in time.

As you may appreciate, smaller sized containers and their mobility platforms catch a ride with larger vehicles by being nested within them. Nesting allows the space in progressively larger vehicles to be used efficiently while the larger size and abilities of these vehicles provide faster transportation at distances farther away. It should also be noted that for containers going to the same location, several containers could share a mobility platform by nesting, rather than each having their own. The individual containers could then be picked up by small mobility platforms for final delivery. While the innovative methods to accomplish this rapidly and efficiently will be covered in subsequent chapters, this option allows for the most efficient use of these transportation components while also respecting available space.

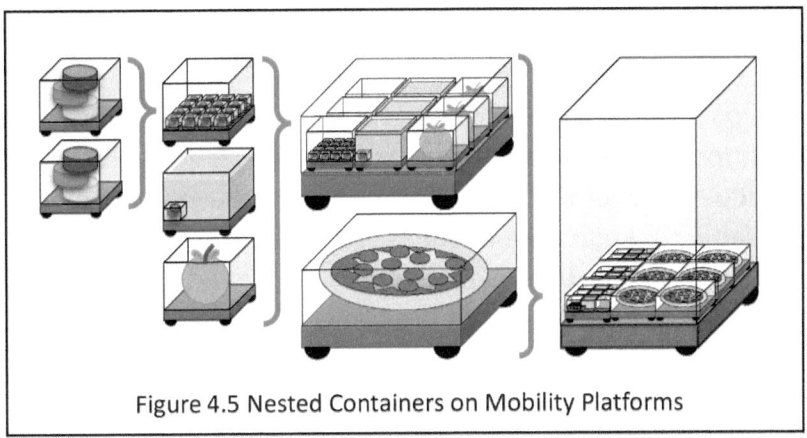

Figure 4.5 Nested Containers on Mobility Platforms

The Scope of Autonomous Vehicles

In talking about bottled water and medications, the use of containers, mobility platforms, and nesting can be readily appreciated. Depending on the contents of the container, the distance of travel required, and the urgency of delivery, various vehicles could be used along the way to facilitate efficiency. Likewise, items of different types and sizes could share rides when the site of delivery was the same or in close proximity. Not only could this pertain to the transport of medications and bottled water, but it could apply to anything...groceries, electronics, furniture, appliances, and even people!

We have described mobility needs, and the benefits of just-in-time medication delivery. With the advancing age of Baby Boomers, these are practical realities where autonomous vehicles serve important needs. But such needs don't stop there. What about the ability to supply a meal to someone who may have difficulty preparing their food? Or the ability to request a particular beverage at a particular time and place? From books to mail to thousands of other items, the opportunities for autonomous vehicles to enhance our lives is tremendous.

This variety of needs leads to a family of automated vehicles matching the needs of the contents and the capabilities of the transportation mode. Our small containers and mobility platforms might be ideal for medications and other items. Perhaps these same vehicles could deliver a piece of jewelry to someone you care about on Valentine's day. Slightly larger sized containers and vehicles would be needed for beverages, meals or books. And even larger ones for more sizable packages like refrigerators and televisions. The beauty of this family of vehicles lies with their ability to nest in one another. Regardless of the size of the items, items with common destinations could travel on a single mobility platform nested within a container large enough to accommodate them all.

Of course, the family of autonomous vehicles is not limited to package deliveries. Certainly, personal mobility vehicles would be included in this family. Other autonomous vehicles could nest within these vehicles when traveling to a destination as well. However, the transport of many people would also need to be considered. By using convoys of larger mobility vehicles, transport of people to any location could be achieved while simultaneously providing transportation to other autonomous vehicles carrying various items. Convoys offer an excellent means to efficiently transport people and packages to various places. Innovative methods of convoy transportation will be presented in a chapter later in the book, but for now, it can be noted that convoys offer much greater efficiency and are much faster than individual vehicles or conventional trains.

Back to feasibility and costs, Tesla has just announced an electric semi for $150,000. This semi is capable of pulling a standard container trailer and includes many of the features needed for self-driving, so those costs are included. A significant fraction of the cost of the semi is the batteries, and as you will see in the next chapter, vehicle batteries can be eliminated because we will power them remotely. With a capacity of 80,000 pounds, this truck is capable of carrying a container with 100 people in their personal mobility vehicles. Thus, autonomous convoy vehicles to carry many people are both feasible and economical.

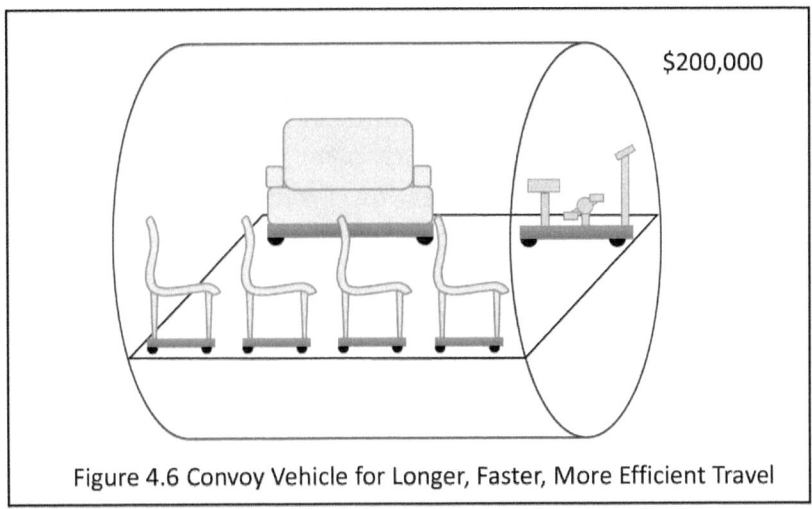

Figure 4.6 Convoy Vehicle for Longer, Faster, More Efficient Travel

This brings up an important point. Remember all that wasted packaging and space we discussed in our current delivery items? One of the important advantages of designing better transportation containers involves their reusability and improved use of space. Certainly, containers being used to transport medications could be returned to the pharmacy or manufacturer for reuse. But at the same time, any packaging materials needed within the container could be reused as well. Even better, containers repeatedly used for the same materials would be designed for those contents, so no packaging is needed at all. All that would be required is container cleaning, which could similarly be automated. Think of the waste this would eliminate! This feature, along with streamlined container and packaging designs, would allow much greater efficiency in the use of both space and transport materials.

While a variety of design types for containers would be important, some standardization would also be needed. For example, each container would need to have a standardized interface to connect to corresponding mobility platforms. A standard locking and unlocking mechanism between containers and mobility platforms is needed for safety and security needs. And a limited number of design sizes, shapes and configurations would aid nesting and transport even further. Establishing some basic standards for containers, creates the opportunity to transport a wide variety of containers for all sorts of items.

Mobility platforms could similarly offer a combination of features involving variety and standardization. From a standardization perspective, mobility platforms would likely share common characteristics such as electrical power interfaces, communication technologies, and latching attachments. These standard features would be needed not only to interface with containers, but with their environment. Mobility platforms would vary in design based on the travel surface, their size, their range, their speed, and their maneuverability. For example, mobility platforms that travel on rails, water, or magnetic levitation are quite specialized requiring greater standardization while mobility platforms that travel on flat surfaces can be more varied. And naturally, larger mobility platforms will have a greater capacity to travel farther and faster carrying larger loads.

Many people today are naturally limited in their mobility. Some are unable to drive or ride in a car, and therefore, they must rely on others or specialized forms of transportation to get them to their destinations. In addition, many people have difficulties simply moving within their homes or in their neighborhoods. My stepmother suffered from such a condition that limited her mobility, and this impacted not only her life but my father's life as well. If only she had access to a functional personal mobility vehicle, she could have continued to enjoy accompanying my father on his walks in addition to many other activities that she otherwise found challenging if not impossible.

Given my stepmother's situation, a personal mobility vehicle that had a container, like a chair, specially designed for her needs would have greatly enhanced her quality of life. With autonomous control, she would not have needed to navigate or negotiate obstacles herself, and at the same time, she would have enjoyed comfort and the benefits of travel. Unlike motorized scooters, which can be quite scary to be around in terms of others' personal safety, an autonomous personal mobility device would be able to detect other people, other vehicles, and various obstacles inside and outside the home. Such a vehicle would have been greatly appreciated by both my stepmom and my father.

You can now better appreciate the opportunities for improving transportation through better vehicle designs alone. By comparing current delivery and transit systems to cargo transportation, the benefits of having standardized containers and mobility platforms are apparent. At the same time, improvements in reusability, mobility, and packaging offer additional advantages. And the use of nesting strategies, as well as matching vehicle types to specific settings, can further enhance transportation quality and efficiency.

In later chapters, we will discuss how such autonomous vehicles will be able to navigate and travel in different environments and the additional benefits possible. But first, we have to establish the basic infrastructure in which these autonomous vehicles will travel. We explore opportunities to design the infrastructure to improve the performance of the vehicles and also to improve the environment. Similarly, some standard features can provide better efficiencies that deserve specific consideration. This will be the focus of the next chapter as we expand further on our concept of autonomous transportation systems for the future of society.

5
Autonomous-Ways (A-Ways): A New Design for Infrastructure

OVER 20 YEARS ago, the two of us sat in a hotel room in Vancouver, British Columbia, preparing for a conference presentation the following day. Our preparation was quite involved, and we didn't want to take the time to leave the hotel for a snack, so we decided to order pizza. Unfortunately, it was after midnight, and finding a place open for delivery seemed impossible. So, we began thinking about how a pizza-size vehicle could bring our pizza to us (even when engineers take "breaks," their minds can't stop designing and analyzing!). By traveling along the ceilings of the halls and the tops of the elevators, the pizza vehicle could eventually deliver our pizza to our room through a slot above the door. That would have solved our immediate problem (hunger) while moving through the existing building with minimal disturbance.

Though our vision of an imaginary pizza delivery vehicle was just for fun, the story highlights the importance of the infrastructure that any transportation system needs. Take our current transportation system, for example. Some infrastructure developments, like interstate highways, had remarkable effects on car and truck transportation.

But at the same time, roadways are exposed to the elements, require expensive maintenance and repair efforts, and consume enormous amounts of space while being limited in scope and speed. And have you noticed that even brand-new roads have bumps? Of course, many of us just accept these limitations because it is the only system we have ever known.

Regardless of the autonomous vehicles we design, they can only be as effective as the infrastructure on which they travel. Therefore, we need to invest some thought about the infrastructure for our autonomous vehicles...an "autonomous-way" (or "a-way" if you will). Fortunately, technology has advanced significantly since our current transportation system was developed over half a century ago, and as a result, these advances allow a great deal of creativity and innovation in designing something more efficient and functional. For example, Amazon is experimenting with drones for package delivery through the air, and Dominos is experimenting with pizza delivery robots that go along the sidewalk, although San Francisco has recently banned delivery robots on most sidewalks. And Elon Musk has proposed transportation tunnels and Hyperloop vehicles in vacuum filled tubes at nearly the speed of sound. Who knows...our vision of pizza delivery vehicles traveling above our heads in buildings may not be so far-fetched after all!

Designing a Better Infrastructure

Imagine for a moment that a total transportation ban was placed on automobiles and buses, and everyone was required to ride the subway or train to get to their destinations. To overcome the inconveniences of this new ban, however, you were selected to provide input about how to improve these public transit systems. Specifically, you were tasked with making them safer, more energy efficient, faster, less expensive, more dependable, less impactful on the environment, more pleasant, and even fun. Where would you begin? While this task seems daunting (and it is!), simply thinking about different alternatives to current approaches to public transit shows the potential for improving our current system.

Let's think about the type of infrastructure that might provide the safest system for our autonomous vehicles...after all, safety first. With this in mind, what makes our current transportation unsafe? As noted, human error contributes to the vast majority of crashes, but we have already addressed this by making our vehicles autonomous. What else tends to make transportation unsafe? Environmental factors and road hazards are among the most common. Weather, potholes, animals, and debris often cause car crashes and injuries. In addition, environmental effects are responsible for wear and tear on roads and highways requiring expensive maintenance and repair efforts as well as for damage to our cars. If we could eliminate these factors and hazards, our new a-way would be a significant improvement over our current transportation system.

In our pursuit to make a-ways safer and more durable, one obvious approach would be to enclose our transportation structures. Think about it. An enclosed structure would protect vehicles from all sorts of weather-related hazards, and at the same time, prevent the possibility of person-driven vehicles, pedestrians, animals, and debris from interfering with the safe and efficient operation of autonomous vehicles. Subways offer a good model in this regard since they are enclosed and utilize electric power supplies, but we can improve on this system tremendously. An enclosed a-way is a strong consideration for autonomous vehicles delivering packages and products since this helps speed delivery while also promoting safety. But these same advantages of enclosed infrastructures can be realized with personal transport as well. Without exposure to snow, wind, rain, dirt, and other environmental factors, safety is naturally enhanced while control and efficiency of transportation is increased.

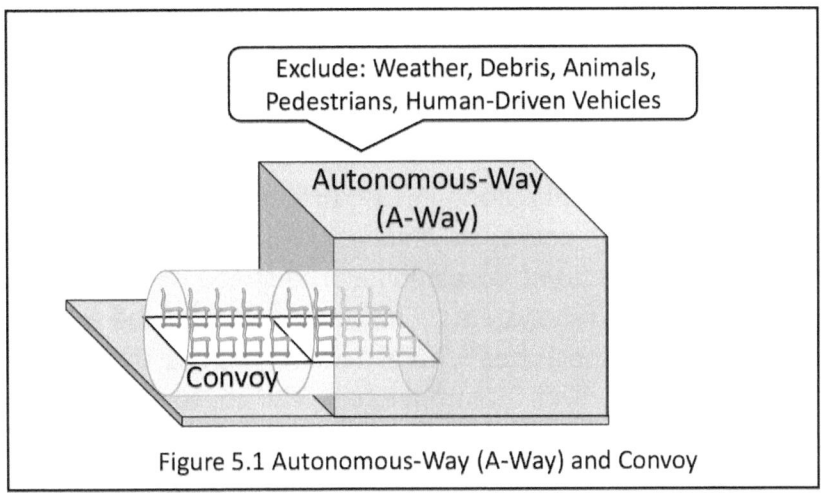

Figure 5.1 Autonomous-Way (A-Way) and Convoy

Enclosed transportation systems have been around for well over a century. Did you know an enclosed pneumatic system was used throughout Paris until the mid-1980s? Over 240 miles of underground tubes (called the "pneu") existed through which mail was sent via compressed air. The system was remarkably efficient and cheap. For less than a couple of dollars, you could send mail anywhere in the city for delivery within 2 hours (and of note, the final phase of delivery was by bicycle).[59] Eventually, the telephone and fax replaced this form of mail delivery transport, but today, banks, hospitals, stores and neutron bombardment facilities still use pneumatic tube delivery structures.

You may similarly be surprised to find out the idea of an enclosed transportation structure for people extends back to the mid nineteenth century. In London, designs and proposals for a pneumatic tube to transport passengers were considered in the 1860s, and similar designs were considered in New York City around the same time

---

[59] Vinocur, John. "Paris pneumatique is now a dead letter." *New York Times*, 1984. Retrieved from http://www.nytimes.com/1984/03/31/style/paris-pneumatique-is-now-a-dead-letter.html

period[60]. While these enclosed tube systems lost appeal in favor of subways, rail systems, and other forms of transportation, the benefits of an enclosed infrastructure were appreciated.

Elon Musk, founder of Tesla and also Space-X, has proposed a similar vision for transportation infrastructures called the Hyperloop System, Musk has encouraged engineers to compete in designing a transportation system that combines the speed of air travel with the convenience of rail systems. The basic design of the Hyperloop System uses enclosed tubes where passenger pods travel to their destinations at speeds of 750 mph! Really. The key to achieving these speeds is the ability to dramatically reduce air resistance and friction, so the use of vacuum and magnetic levitation technologies has been suggested.[61] But once again, the success of such a system utilizes the benefits of enclosed infrastructures. By shutting out external interference, transportation efficiency and speed can be greatly enhanced.

Musk's Hyperloop System has some potential challenges, such as the need for extremely straight and flat tubes that extend long distances to achieve the high speeds proposed. But the concept highlights another strong advantage of enclosed infrastructures. Without environmental and external influences, road surfaces within these structures can be designed to best match transportation needs. In the Hyperloop System, the use of magnetics to minimize the impact of a solid road surface is being considered. This might be ideal in some situations, but other a-ways might be best designed by having other types of road surfaces for shorter distances or lower speeds. Elon Musk has formed The Boring Company to drill tunnels to create just such enclosed transport environments.

---

60 Campbell-Dollaghen, Kelsey. "How one inventor secretly built a pneumatic subway under NYC." Gizmodo.com, 2013. Retrieved from http://gizmodo.com/how-one-inventor-secretly-built-a-pneumatic-subway-unde-1123695775

61 Irving, Michael. "Stranger than fiction: Inside the SpaceX Hyperloop Pod Competition." NewAtlas.com, 2017. Retrieved from http://newatlas.com/spacex-hyperloop-pod-competition-vichyper/48252/

Do you remember the description of the virtuous cycle involving autonomous vehicle design? As safety measures were improved, and energy efficiency enhanced, the weight of the vehicle could be further reduced leading to additional benefits in safety and energy efficiency. The interplay between a-way road surfaces and autonomous vehicles extends the virtuous cycle. Smooth road surfaces facilitate faster speeds and eliminate heavy tires and suspensions, and lighter weight vehicles have less of an impact on road surfaces. We have the opportunity to optimize the a-way surface to match the autonomous vehicle, whether it is a flat surface, rails, or even magnetic or air levitation. As each is designed to be more efficient and effective, the other gains the potential for additional improvements.

Safety and efficiency advantages are not the only benefits of enclosed a-ways. For instance, an enclosed structure would enhance the speed and reliability of transportation communication systems. From an information management perspective, enclosures offer protected environments for transportation management and control systems. And for customer communications, several enhancements can be enjoyed as a result as well. Mobile communications systems, Wi-Fi networks, and even optical communications systems can be better constructed within a-ways to provide higher quality services with enhanced bandwidths. With this in mind, enclosed a-ways would offer protected communications for autonomous vehicles as well as for their passengers and packages. Speed, location, and a host of other communications data could be readily accessed and transferred through the a-way infrastructure environment. And better communication routinely promotes better safety, efficiency and quality.

How might we make a new transportation infrastructure more energy efficient? As discussed, gasoline powered automobiles are highly inefficient, and this is even more apparent when we compare them to freight trains and electric vehicles. But even freight trains and electric cars have their issues. Electric cars are energy efficient, but they require large electric batteries to operate. Batteries add significant weight to a vehicle and utilize lithium, which is a scarce resource. In terms of freight trains, many operate on diesel fuel, which is not environmentally friendly. Neither of these options are ideal.

One of the best things about an a-way is its ability to allow power to be supplied to vehicles from an external source. Rather than having a vehicle with its own big battery, power could be supplied directly to all vehicles in a-ways, as happens with subways and electric trains today. Our vehicles would be much lighter while enjoying a more efficient source of energy. Later chapters will describe innovative techniques to further increase both speed and energy efficiency in this area well as others.

With a-ways delivering electrical energy for vehicle travel, there would not be any need for battery recharging stations. This would reduce the need for additional infrastructures as a result. Likewise, the benefits of regenerative braking that were mentioned previously are enhanced because energy recovered from one vehicle braking can be transferred to a nearby vehicle accelerating or running. And by being connected to the external power grid, system shutdowns due to energy shortages would also be remarkably less likely since power grids inherently have built in redundancies and better reliability. A-Ways certainly offer many advantages when it comes to energy efficiency.

Lastly, we need to have an infrastructure that is environmentally friendly. By using electricity as an energy source, pollution from the vehicles is minimized and renewable resources can be used to generate the electrical power, even on the roof of the a-way. An enclosed structure also eliminates water pollution since vehicle and road debris in water runoff would no longer be present. At the same time, enclosed structures significantly reduce noise pollution, and because they have higher capacities, a smaller infrastructure could be designed to be much more aesthetically pleasing. Ironically one of the complaints about electric cars is that they are too quiet, so people can't hear them coming — a-ways eliminate that problem and contain even that low noise. Think about all those homes, offices,

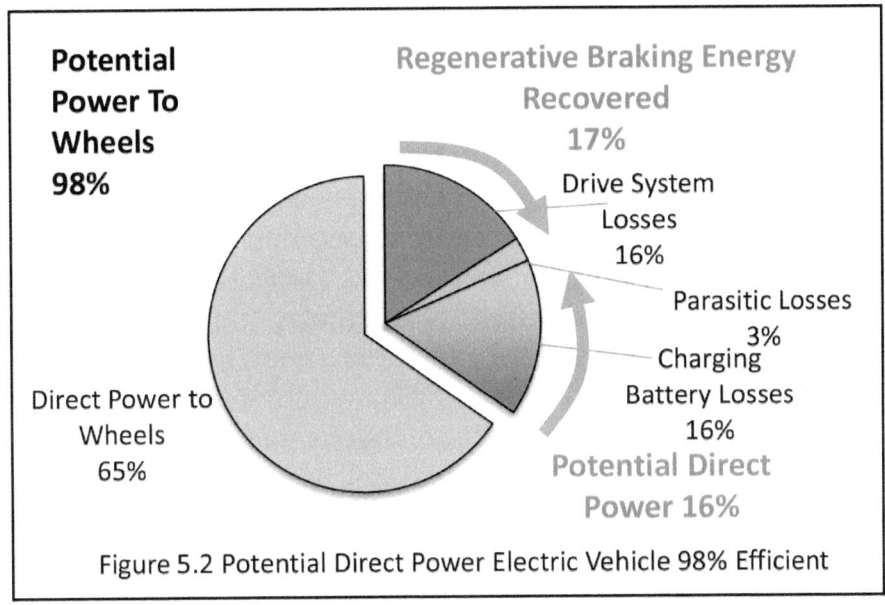

Figure 5.2 Potential Direct Power Electric Vehicle 98% Efficient

warehouses, and parks currently situated near railways and highways today along with the associated noise and debris. A newly designed, enclosed transportation infrastructure could make these nuisances a thing of the past. And in the process, we gain cleaner air, cleaner water, and an environment much more satisfying.

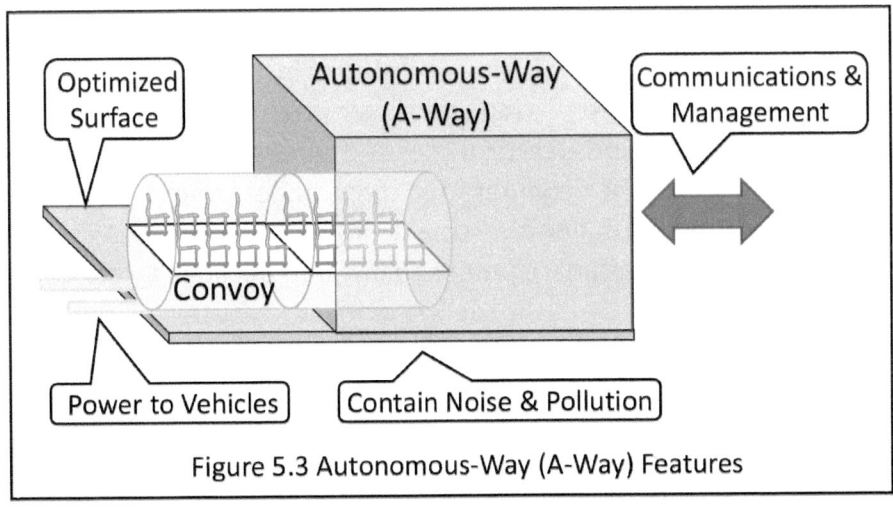

Figure 5.3 Autonomous-Way (A-Way) Features

A-Ways Offer Tremendous Advantages Beyond Transportation
We have already shared the notable inconveniences associated with road repairs and the related infrastructure when describing summers at Tupper Lake. But these are relatively minor when talking about the economic costs associated with these never-ending repairs. Just think about the work crews, heavy equipment, and business disruptions that occur every time an asphalt road has to be dug up to replace buried pipes or wires, or to fix potholes. It is easy to appreciate these costs are substantial, but at the same time, delaying such repairs is even more costly.

For example, In Los Angeles currently, twenty percent of the city's water pipes were installed before 1931, and nearly all will be outdated within another decade or so. The estimated cost to replace these pipes exceeds $1 billion, and this fails to account for business interruptions and other problems associated with these repairs. However, choosing to ignore the problem is much worse. On average, the city sends out water repair crews to 4 pipe leaks a day that cause billions of gallons of water to be lost. In a year's time, the amount of lost water from these leaks could provide water service to more than 50,000 homes![62]

You might assume that placing such structures above ground might offer a better alternative. After all, we are sure you have seen the various colors of spray paint on grass and sidewalks identifying underground infrastructures that are placed by city workers to prevent damage from construction and repair projects that involve digging. And above-ground electrical lines and wires are significantly less expensive to place when compared to those underground. But above ground lines demand major investments in repairs over time. The city of Washington D.C. estimated more than 1,000 outages a year could

---

62 Poston, Ben, and Matt Stevens. "L.A.'s aging water pipes; A $1-billion dilemma." Los Angeles Times, 2015. Retrieved from http://graphics.latimes.com/la-aging-water-infrastructure/

be prevented simply by placing such structures underground.[63] We have watched in distress as crews from telephone, cable, and power companies spent an entire day replacing a single utility pole knocked down by a drunk driver while two police officers directed traffic.

Here again, a-ways offer a tremendous potential to improve not only our lives but also the financial bottom line. What if all these infrastructures could be housed within our enclosed a-ways? Think about it. There would be no poles or painted ground markings. Digging and cutting up roads and terrain would be avoided. And at the same time, everything would be enclosed and protected from environmental effects. Such a system makes much more sense than the systems we have in place today.

Of course, the benefits of enclosing other infrastructures within a-ways do not stop there. What if a repair to one of these infrastructures was needed? By designing a-ways to have greater accessibility to each infrastructure without the hassle of damaging the existing structures, repair crews could easily locate and repair any problem efficiently without disruptions in transportation or other infrastructures. Likewise, the repairs could be performed in a fraction of the amount of time that is required currently saving additional costs in labor and lost business operations. Similarly, installation and upgrades are cheaper and not disruptive.

Let's take this a step further. By designing automated installation and maintenance systems within our a-ways, the need for repairs would be reduced over time, and problems would be more likely to be corrected before they got to a critical state. These types of automated systems exist today in many industries, and the economic benefits of prevention and maintenance are clear. Not only would installation and repair costs decline, but the labor associated with maintenance efforts would also be dramatically less. The economic advantages of these approaches can be easily appreciated from a common-sense perspective.

---

63 Plumer, Brad. "Why most cities don't bury power lines." The Washington Post, 2012. Retrieved from https://www.washingtonpost.com/news/wonk/wp/2012/07/25/why-most-cities-dont-bury-power-lines/?utm_term=.73a34e91cda2

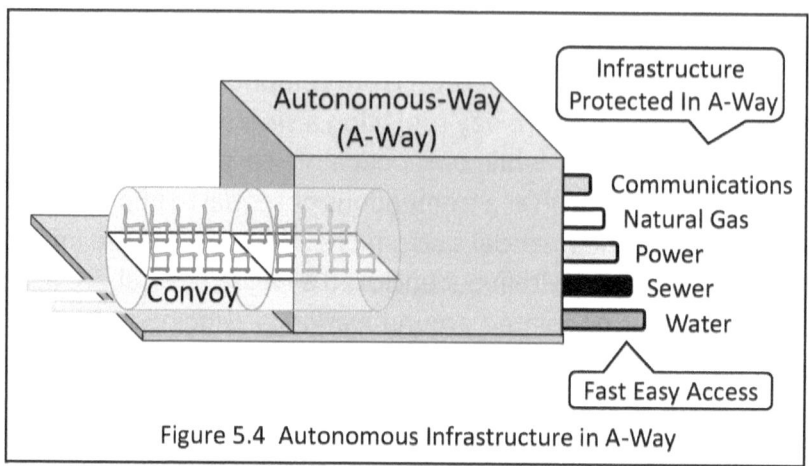

Figure 5.4 Autonomous Infrastructure in A-Way

From a transportation standpoint, a-ways offer many benefits over our current systems. These alone supports the adoption of such a system since the costs in lives, time, and money we currently incur from transportation is enormous. The economic burdens related to energy inefficiency, poor safety, and negative environmental impacts more than substantiate the need for a better transportation model. But if we also consider the additional advantages a-ways offer in relation to the housing of other infrastructures, these financial savings become compelling. Not only would a-ways pay for themselves in a short amount of time, but they would also enhance our economic position overall for decades to come.

Matching A-Ways to Autonomous Vehicles

Having an enclosed a-way solves many problems as we have described. Road hazards, pedestrians, stray animals, and weather issues are avoided while communications, control and maintenance are enhanced. Likewise, our enclosed a-ways provide the ability to supply electrical power to autonomous vehicles while also permitting the enclosure of other non-transportation infrastructures. However, we have yet to discuss how specific a-way designs can best match specific autonomous vehicles.

Remember our small autonomous medication container and its mobility platform? Given its size and its need to travel through pharmacies and homes, it would need to travel inside many different locations to get medication to patients just in time. Understanding this, some autonomous vehicles will need to be capable of traveling indoors while others will not. In other words, some will need to be able to maneuver freely and independently within an open area such as one's home or office while others may be confined to a-ways only.

Within a building, the medication vehicle must maneuver around walls and through doors while avoiding people, pets, potted plants, and other vehicles. It must safely and efficiently navigate to the person in need of the medication, once it arrived within the vicinity of that person's home. It needs to move on a variety of floor surfaces including carpets, and its sensors and control mechanisms must deal with the complexities of the surrounding environment, perhaps calling for a different mobility vehicle. For example, moving safely around people requires a lower speed. These design constraints would apply to any autonomous vehicle that needed to move around in these open spaces. However, a-ways can also be installed inside buildings to avoid these challenges.

Other autonomous vehicles may never leave an a-way. For example, larger autonomous vehicles that carry the medication containers and other vehicles across town or across the country may be permanently within enclosed a-way structures. Because of this, these vehicles would not require the same sensors and control mechanisms to navigate open spaces as indoor vehicles would. Vehicles that did not need to navigate indoors or in the open could be designed to focus on speed, capacity and range through the use of different sensors and control mechanisms. We will expand on these differing capabilities in future chapters. Ultimately, the function of the autonomous vehicle, and its need to be operate indoors or not, would influence its design and construction.

So how might such a system of a-ways and indoor vehicles function together? Let's consider the analogy of the body's circulation system. Within each of our bodies, we have arteries and veins. Arteries carry oxygen and other nutrients to our body's tissues while veins remove carbon dioxide and waste products from these same tissues. But not all arteries are the same, and neither are all veins. Some are larger allowing greater volumes of blood to flow rapidly to or from the heart. Others are small allowing blood flow to slow down as they travel deeper into our tissues. The different sizes of these structures have a great deal to do with the functions they serve, and a-ways can be considered in the same way. And once the blood reaches the tissues, it travels outside both the arteries and the veins. Thus, we have vehicles that travel at different speeds in corresponding a-ways, and some vehicles that can travel outside of a-ways.

To create an efficient transportation system, an infrastructure must match the purpose of its vehicles. Small autonomous vehicles need to negotiate small spaces while traveling at slower speeds. In contrast, larger autonomous vehicles will need to cover greater distances as quickly as possible. Therefore, like our bodies' circulation system, a-ways need to be constructed to accommodate the intended function of different sized vehicles. While a Hyperloop System may be perfect for getting vehicles and people across the state or country, other types of a-ways will be needed inside buildings. Size, range and speed of the autonomous vehicles traveling a particular a-way will determine how it should be best configured.

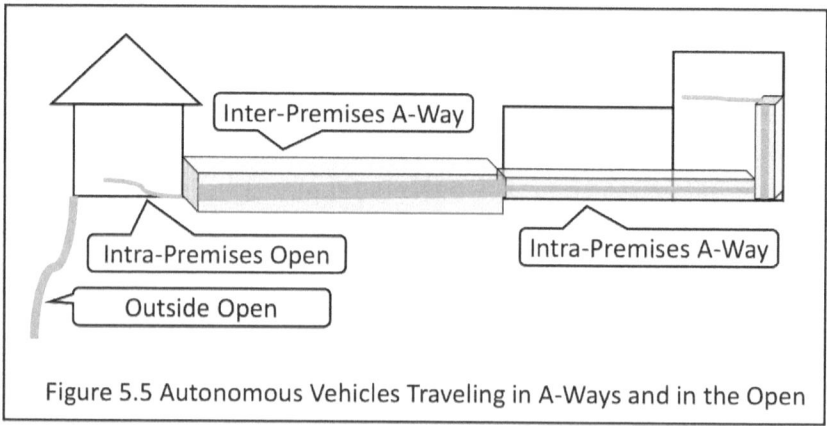

Figure 5.5 Autonomous Vehicles Traveling in A-Ways and in the Open

Negotiating Our Current Landscape

Wouldn't it be nice if we could construct our new transportation system without any constraints? Unfortunately, we have an existing infrastructure in place as well as many other obstacles that require our attention before we can just go putting up new a-ways. For example, how do our new a-ways get around existing roads and highways? How will they be situated in relation to existing homes and buildings? Should they be constructed above the ground or below? All of these are relevant questions when thinking about the implementation of our new transportation system. And while the exact answers will vary depending on specific situations, we can consider some of the possibilities given the existing transportation and infrastructure environment.

Have you ever been to Amsterdam? If you have, then you know the city loves its bicycles. Did you know there are over 800,000 bicycles in Amsterdam alone? By comparison, fewer than 300,000 cars exist. In fact, nearly two-thirds of the residents use their bicycles daily, and more distance is traveled by bicycle than by car.[64] While this speaks to many benefits in terms of environmental sustainability, bicycles are certainly less than ideal when inclement weather hits (another example of how an enclosed system would be beneficial). Despite this, Amsterdam offers an example of how two different transportation infrastructures can be designed to coexist.

In implementing a new transportation system, existing structures have to be considered. Let's consider smaller a-ways and the obstacles they might encounter. Remember our imagined pizza delivery vehicle? If we had the luxury of constructing a new building, then we might place an a-way path through the walls, ceilings or floors to reach each room, so it would not interfere with any other structures or activities. But this may not be the case for existing buildings. If

---

64 Van der zee, Renate. "How Amsterdam became the bicycle capital of the world." The Guardian, 2015. Retrieved from https://www.theguardian.com/cities/2015/may/05/amsterdam-bicycle-capital-world-transport-cycling-kindermoord

adequate space in the ceilings or floors does not exist, then other options for a-way paths must be considered. One possibility might be constructing a-ways along the walls, either inside the building or attached to the outside, to facilitate unobstructed travel of small vehicles from place to place.

Naturally, a-ways will vary in their size depending on the size of the vehicles they transport. For example, if medications or jewelry were primarily being transported, an a-way could be 2" square, which would easily fit inside walls, ceilings, or in the corners of hallways of existing buildings. For transporting food, beverages, mail, and flowers a one-foot square a-way might be sufficient, which could be incorporated in a new building design, within the dropped ceiling of an existing building, or attached to the outside of the building. Using nesting techniques, one of these larger containers might carry over 100 medication containers. For the transportation of people and appliances, an a-way might be 6' square, which might be placed in the basement or on the roof of existing buildings. A container designed for this size a-way might carry over 100 food containers (or over 10,000 medication containers). A-way design will align with the functional needs of the transportation needed including vehicle volumes and sizes.

In providing transportation between buildings, different issues may need to be considered. Let's assume our small a-way has connected to a medium size a-way as it reaches the perimeter of the building. What then? Once again, different options might exist depending on the situation. In some cases, medium size a-ways might travel alongside existing roads just like the bicycle paths in Amsterdam, perhaps in a median if it exists. Alternatively, we might suspend them like a monorail. In fact, some of the bicycle paths in Amsterdam are elevated in this fashion. A-ways might travel along an existing roadway, attached to buildings, or elsewhere. Or we might place them underground if this offers the best choice. These determinations will certainly vary depending on existing situations.

The most common issue in dealing with congested urban environments involves private property ownership of land and structures. In constructing a medium or larger size a-way, paths must be negotiated based on existing structures as well as on the presence of privately-owned properties. While eminent domain options might exist, such strategies are rarely welcomed by the community or easily approved. For this reason, developing a-ways along existing roadways and public land offers some natural advantages.

One alternative idea involves creating a network of tunnels beneath the surface. In addition to Tesla and Space-X, Elon Musk has also started a company called "The Boring Company," which is seeking to bore a network of tunnels to accommodate high-speed underground travel. Musk's vision calls for platforms (called "skates") along existing roads where vehicles will enter and park in these skates — in our terms, the skates are mobility platforms, and the cars are containers, or autonomous vehicles nested on the skates. The skates then drop several stories below ground where the skates (along with their cars) merge onto high-speed tracks allowing travel speeds up to 125 mph. With easy merging lanes, automated controls, and programmed destinations, the system is highly efficient and practical.[65] A major advantage of tunneling is avoiding private property issues!

A-ways offer many opportunities for enhanced efficiency. In addition to the efficiencies enjoyed from lighter weight vehicles and external electrical power supplies, a-ways would permit the ability to house many other types of infrastructures that would be protected from the environment while offering easy access for maintenance and repairs. Likewise, through the use of vehicle convoys (which will be discussed in detail in later chapters), the ability to efficiently use space will be greatly improved. Given these advantages, a-ways could offer 15 times greater transportation capacity as well as 24 times greater efficiency when compared to cars on a highway, and that is for carrying people – smaller loads gain a much larger advantage

---

65 Mack, Eric. "Elon Musk shows how his Boring Company plans to tunnel under traffic." NewAtlas.com, 2017. Retrieved from http://newatlas.com/elon-musk-boring-company-tunnel-concept-traffic-spacex/49283/?li_source=LI&li_medium=default-widget

due to nesting. Just imagine the impact this would have if such a system of a-ways could be developed beneath the surface allowing the transport of all kinds of vehicles, goods and passengers!

The decision for the design and construction of a-ways will thus be very dependent on the existing environment. This not only pertains to the geography and existing physical structures but also on available economic resources and social policies.

While our current transportation system does impose some constraints on the development of a-ways and a new infrastructure system, these constraints are not complete roadblocks by any means. New building construction and a-way development certainly offer great opportunities to enhance our networks of transportation. But at the same time, innovative concepts can be applied to existing roads, highways, buildings and environments to effectively accommodate a-ways and enhanced travel. In the chapter on Linear Cities we will see how the a-ways can be incorporated into the fabric of a city to provide even further benefits and to create an irresistible economic advantage.

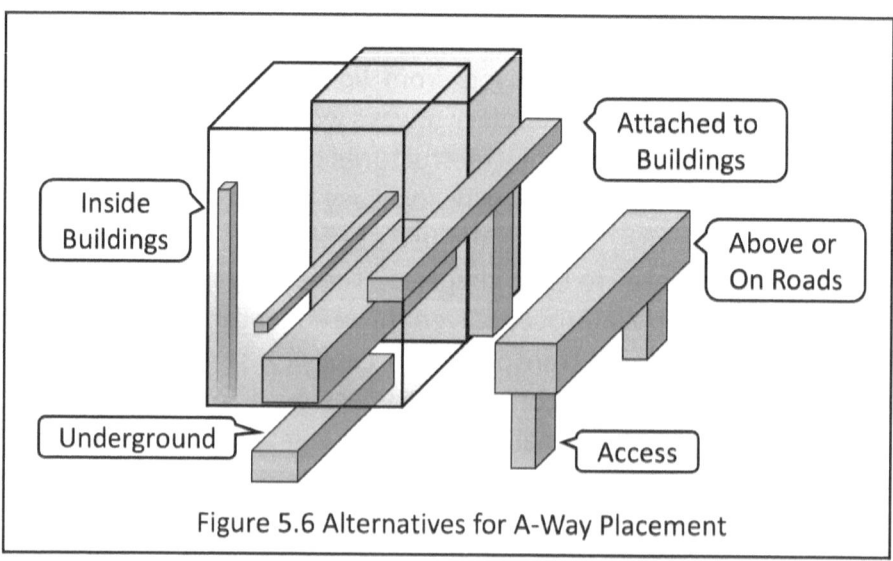

Figure 5.6 Alternatives for A-Way Placement

Do you recall the comparison we made between the efficiency of a freight train carrying a pound of oranges versus a car? In essence, the train offered enormous advantages in energy use traveling 30,400 times as far on a tablespoon of oil when compared to an automobile — the incredible magnitude of this advantage emphasizes both the opportunity for improvement, and the inevitability of such changes. This simple fact alone highlights the amount of waste within our current transportation system. While this may be quite depressing at first glance, a silver lining also exists. By dramatically reducing this waste, the savings can be used to design and develop an entirely new transportation network! And with the enormous savings gained in installation, maintenance and repair benefits from having other infrastructures incorporated into these a-ways, our entire transportation system could be funded from these improvements and savings, especially for new construction.

With lighter weight autonomous vehicles, the design and construction of a-ways offer major improvements in capacity, efficiency, and costs when compared to our current roadway systems. By enclosing these structures, we enjoy many other advantages in safety, maintenance, control, and communications. And by utilizing these a-ways to supply electrical power and the housing of other non-transportation infrastructures, the benefits are greatly multiplied. While constructing new a-ways for our autonomous vehicles offers some short-term challenges financially and otherwise, incentives are there to not only push these ideas forward but to also demand our attention for the future. The positives of an a-way system and autonomous vehicles far outweigh the negatives. In fact, it is not even close.

In this chapter, some key considerations in a-way infrastructure design and development have been considered. Likewise, some innovative approaches to new transportation system concepts have been highlighted. There are many other innovative proposals coming in future chapters that also offer tremendous advantages. A-way access ports and "autonomous doors" to provide privacy and security are a couple of these ideas, while autonomous monitoring, testing, and repair provide greater assurances in the safe and efficient operation of the system. While some of these concepts are likely to be universally adopted (such as enclosed structures), others may vary depending on specific settings and environments. Regardless, realistic and practical solutions for an advanced and efficient transportation system do exist. And based on the inefficiencies and costs of our current system, the time is now for such solutions to be pursued.

6
Continuous Convoys

WARREN: LET ME share my frustrations with commuter trains. On many occasions, I have taken the train into Manhattan, and each time I became increasingly bothered by the long travel times. On the regular local commuter train, frequent stops added at least a half hour to the trip, and even the express train took over an hour to travel 50 miles. I was further frustrated with the lack of scheduling options. During the day, the train ran once an hour with a few more times during peak rush hours. But at night, no train options were available. And weather delays, mechanical problems related to poor track conditions, and traffic congestion contributed to additional delays as well. In my mind, I knew there had to be a better way.

Many of my frustrations have been addressed with the presentation of autonomous vehicles and a-ways, and this chapter seeks to introduce an additional way to enhance transportation even more. We will discuss how autonomous vehicles will traverse a-ways by presenting the concept of continuous convoys, which offer many advantages over current transportation models. Inherent to this subject are additional considerations that will be discussed as well, like en-route sequencing and stacking. And a realistic scenario of how these concepts can be applied will be provided. Not only will you better understand how such a system can help alleviate my own irritations regarding our transportation systems currently, and save energy, but you will also come to appreciate how much better they can be.

Continuous Convoys

When you think about public transportation, certain issues immediately come to mind. How will you get to the station? If you drive there, where will you park? If you take a taxi, how much will it cost? Then, of course, once you arrive at the station, you are confronted with other dilemmas. When will the next bus, train, or shuttle arrive? If you have a connection, how long will you be waiting there? And once you are on board, you must deal with frequent stops along the way, delaying the time it takes to get you to where you are going. An express train may not even stop at your desired station.

These are some of the more obvious problems associated with modern public transportation, but at the same time, it does have some benefits. Taking public transportation often costs less than taking your own vehicle, especially in crowded urban areas. Likewise, once you are traveling, you have time to read, check social media, or other personal activities that you otherwise would not be able to do (or should not do) if driving. And public transportation contributes less to noise and air pollution than cars and requires less energy per passenger for transportation activities. The key would be to design a better system that enjoyed these benefits yet corrected the other major problems commonly associated with this form of travel.

This is where continuous convoys come into play. What is a continuous convoy? Like subways and trains, continuous convoys consist of a series of connected vehicles that travel together. Having already discussed autonomous vehicles and a-ways, envision a series of large autonomous vehicles (called convoy vehicles) connected together and traveling down an a-way (no different than multiple cars on a subway train). But unlike subways and passenger trains, our continuous convoys don't stop and start at every station. This allows them to move "continuously" along the a-way eliminating the inherent delays we are used to when using other forms of public transportation.

You are likely asking, "How is that possible?" In order to answer this question, let's first take a step back and describe what an actual major a-way might look like. You may be expecting crowding next to a track with autonomous vehicles racing by or coming to a halt in front of you. If so, you are in for a surprise. You begin by planning your trip on your electronic assistant. You see a series of doors ahead, similar to elevator doors. Your electronic assistant indicates which door to enter. Just as you arrive the door opens and you step in, along with other people.

Within a few seconds the doors close and you feel yourself descending. Within a few more seconds you stop descending and start to accelerate. The ride is smooth and quiet. Your electronic assistant may advise you to walk through a door to another vehicle. There are seats along the sides of the vehicles, but plenty of room to walk, or ride your personal mobility vehicle. If you are taking a longer trip, you may feel your vehicle slow down and shift sideways into a different lane, then accelerate again. In a few minutes your vehicle slows, stops, and you feel it rising. The doors open and you have arrived at your destination. You never see speeding vehicles or empty tracks, no loud noises or blowing dirt, no rushing crowds.

What you haven't seen is the infrastructure that makes all this possible. The space you entered through the doors was a container that was lowered onto a mobility platform for the appropriate speed and direction of your travel. What happens next depends on where you enter the a-way and the next station. For example, every mile your vehicle can join a 30-mph lane of the a-way. If you weren't close to one of these stations, an elevator can take you there. But this is unlike any elevator you have seen, as we will describe in the next chapter. Likewise, your vehicle could then access the next adjacent 130-mph lane every 16 miles. And finally, you could then access the center 260-mph lane every 64 miles. You might even be lucky enough to have access to a 520-mph lane or a Hyperloop lane for incredibly fast travel speeds. Depending on your destination, the distance you were traveling, and the speed with which you wanted to arrive, your vehicle would access the appropriate a-way lanes needed to make your trip as efficient as possible.

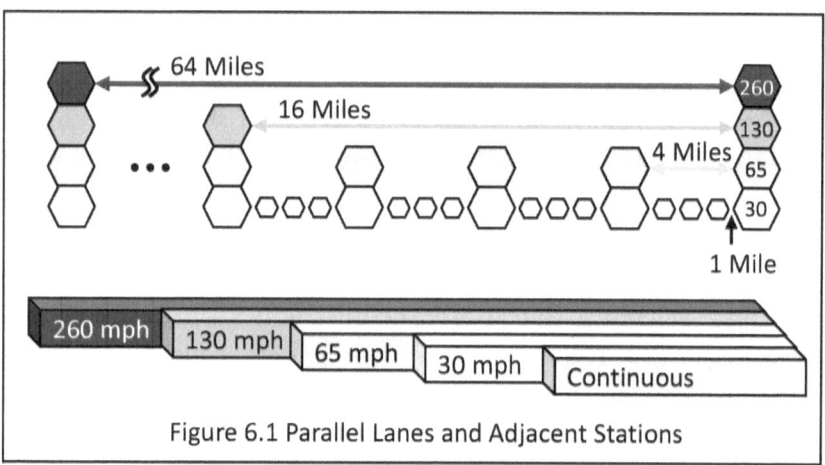

Figure 6.1 Parallel Lanes and Adjacent Stations

By allowing different speed lanes within the a-ways, transportation can be better aligned to actual travel needs. Of course, this applies not only to individuals but to all sorts of items. From people to packages, the best combination of speed lanes in the a-way would be used to arrive at a destination as fast as possible. With access to each speed lane being quite frequent, waiting delays are also kept to a minimum, as you will see shortly. In fact, you could be going 130 mph in less than 10 minutes. That's quite an improvement over our current public transportation systems! For example, it would take longer just walking to a station and waiting to board a bus or subway. In 40 minutes, you have traveled between 85 and 250 miles, depending on where you start, and you are traveling 520 mph. It would take you longer than that just to get to an airport and board an airplane.

Let's now consider how our convoy of autonomous vehicles would be able to move continuously without stopping. Have you ever seen a relay race? Relay races are track races where the first runner in the medley passes a baton onto a second runner, who then races to pass the baton onto a third runner, and so on, until the race is complete. If you have ever watched such a race, as the runner with the baton approaches the next runner, the runner receiving the baton accelerates so the transfer of the baton occurs while in motion. By matching their speeds, transfers can be made more safely while the overall race times can be as fast as possible.

We can apply this same concept to our continuous convoys. Imagine we have a convoy of four large convoy vehicles connected to one another. For the sake of identifying them, let's label the convoy vehicle in front as "A" and the last one as "D" with the intervening convoy vehicles being "B" and "C." Now, let's assume our convoy is traveling along an a-way at 65 mph, and an a-way station is up ahead. Some of the passengers and items in the convoy want to exit at this station, but we don't want to stop the entire convoy for this purpose. In order to do this, we position all the items and passengers that would like to exit at the station in the last convoy vehicle, labelled "D." Just before the convoy reaches the a-way station, vehicle "D" detaches itself from the continuous convoy and begins to decelerate so it alone can stop at the station. Meanwhile, the rest of the convoy (convoy vehicles "A" through "C") continue along at 65 mph without any delay.

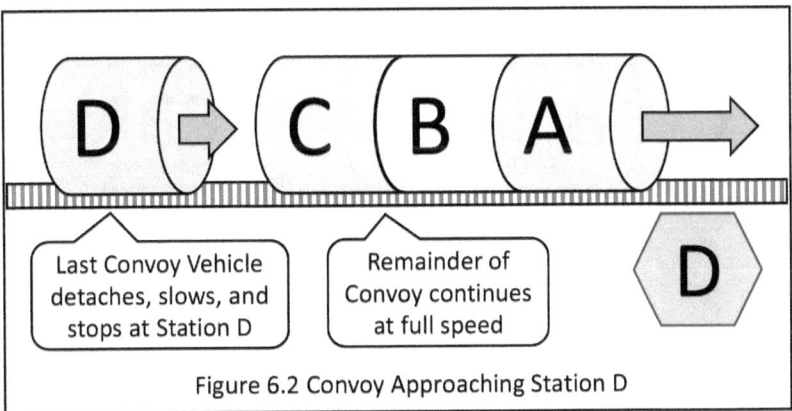

Figure 6.2 Convoy Approaching Station D

This type of design completely eliminates the multiple stops and starts we now encounter when traveling on many public transportation systems today. By allowing only the convoy vehicle containing people and items wishing to exit at a particular stop to detach from the convoy, the remaining convoy vehicles can continue forward at full speed. Think how much faster your commute or trip might be now if these unnecessary delays were eliminated! During rush hour, the avoidance of these delays could make a substantial difference in the time it took you to get to your destination.

What if you wanted to join a continuous convoy rather than exit one? Suppose you are at an a-way access station in your personal mobility vehicle. You (or your programmed autonomous vehicle) determines which large convoy vehicle at the station to enter, and you then load into the convoy vehicle and await the oncoming continuous convoy. As the continuous convoy that you wish to join approaches the station, your large convoy vehicle begins to accelerate (just like our runner receiving the baton). By the time the convoy actually catches up, your convoy vehicle has matched the speed of the overall convoy and joins the convoy by attaching to the front. Because the speeds are identical by the time the continuous convoy arrived, your convoy vehicle is able to join the continuous convoy without causing any reductions in the overall travel speed.

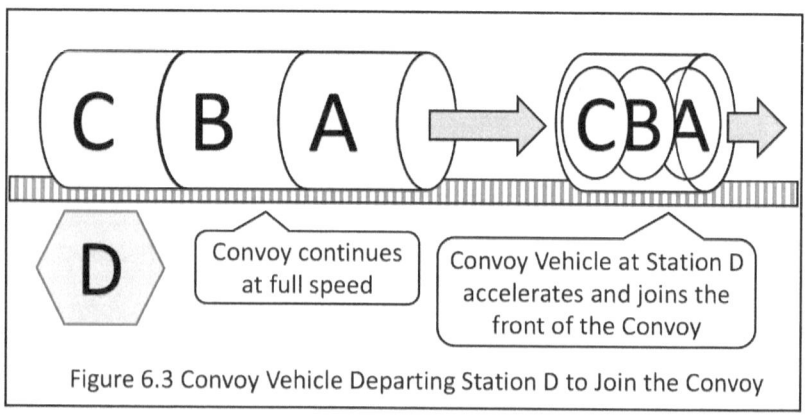

Figure 6.3 Convoy Vehicle Departing Station D to Join the Convoy

You might be wondering what happens at a major station, such as Philadelphia's Penn Station, where a lot of people want to get off at the same time. The continuous convoy handles this case easily — several convoy vehicles can detach and stop at the station, even a majority of the convoy. Similarly, many convoy vehicles can join at the front of the convoy. Thus, people continuing on toward New York City do not need to stop, even though many people get off in Philadelphia, and many get on.

As you can appreciate, this type of transportation design addresses many of the problems and delays associated with current public systems. By having autonomous vehicles, you no longer need to park near a public transportation station. Instead, your autonomous vehicle simply accesses the a-way itself. And through the use of continuous convoys, you no longer are bothered with frequent stops and starts on the way to your destination. Such a design has clear practical advantages over modern public transportation models.

En Route Sequencing

In our previous example describing continuous convoys, we showed how convoy vehicles joining a continuous convoy attach in front while exiting vehicles detach from the back. This allows continuous speeds without convoys having to stop and start, and it still allows safe and efficient entering and exiting from the convoy. But how do you (or a package) get from the front to the back of a convoy before you arrive at your destination? Does your personal autonomous vehicle automatically know to do this, or is this something that you have to navigate yourself?

The process by which this occurs within continuous convoys is called en-route sequencing. Though you may not be familiar with this process, en-route sequencing is actually not a new idea. In fact, the U.S. railway mail services used en-route sequencing techniques as early as the mid-nineteenth century as a way to speed up mail delivery times. As trains would travel along a route, workers would sort the mail according to its final destination. Once the train arrived at the second station, the mail was already organized for local delivery making the entire process much more efficient. In other words, mail was "sequenced" while the train was "en-route" so that the mail was ready for delivery by the time the railway arrived at the station. In some cases the train didn't even stop, the mail bag was grabbed by a mechanical arm while the train continued.

Let's apply that same concept to our continuous convoys. As you join a continuous convoy, based on your final destination, you can move to the correct convoy vehicle that will be exiting at that location. Therefore, after joining the convoy, you will need to move through the convoy to the correct convoy vehicle. If you are exiting at the next station, then you will move to the last convoy vehicle in the convoy, since this will be the one that will detach, decelerate, and stop at the next station. If you are exiting a few stations ahead, you will move to that particular convoy vehicle. In order for our continuous convoys to remain "continuous," en-route sequencing is essential.

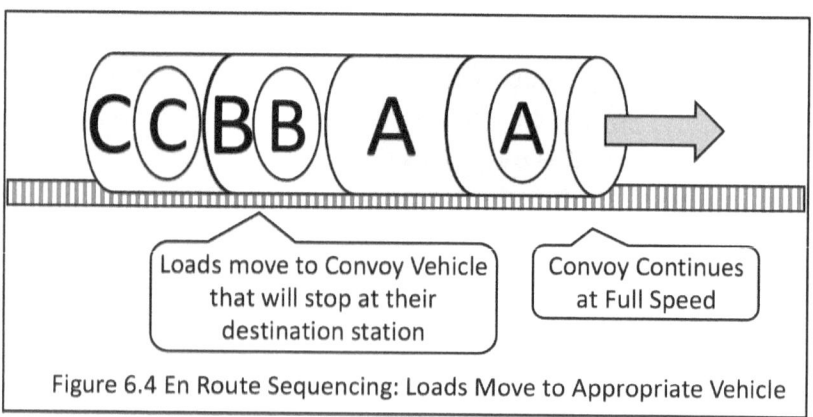

Figure 6.4 En Route Sequencing: Loads Move to Appropriate Vehicle

How do you get from one convoy vehicle to another once you are traveling on the continuous convoy? One option would be to program your personal mobility vehicle or autonomous vehicle to automatically move to the correct convoy vehicle on its own. By entering your desired destination into an interactive scheduling operating system at the beginning of your trip, your vehicle would determine your route as well as your en-route sequencing that would get you to your destination as efficiently as possible. At the same time, you may simply load into an autonomous vehicle originally that aligns with your desired destination. Once it joins the continuous convoy, this autonomous vehicle will move to the appropriate convoy vehicle to ensure you get to where you want to go.

What if you wanted to travel by foot instead of traveling in an autonomous vehicle? Exercise is certainly an important part of a healthy life, so the entire transportation system should encourage walking and other forms of exercise. Today we use map apps on our smartphones, a similar apps will direct you to the correct convoy vehicle. Of course, you need to arrive before that vehicle detaches from the convoy. Have you ever watched frequent travelers on the subway or train? They move about without thinking about it, even reading the newspaper or their smart device. The ability to mix walkers with people on personal mobility devices is an important design goal, as we discussed with my father and step-mother.

Considering all the en-route sequencing activities occurring during continuous convoy travel, you are probably wondering what a convoy vehicle might look like. While the exact design may vary based on loads, and design preferences, some features can be assumed to be present based on the required components of such a transportation system. For one thing, the size of these convoy vehicles must be able to accommodate its contents as well as en-route sequencing activities. Depending on the volume of people and packages traveling in a given area at a given time, the actual size of a convoy might vary as might the size of the individual convoy vehicles. Note, for example, we could have convoy vehicles and corresponding a-ways designed just to deliver lunches and beverages, so these would be much smaller than those designed to carry people. Regardless, space and size considerations will be important so that all transportation items can be well accommodated.

Likewise, convoy vehicles must be able to interact effectively with each other to carry out the different actions required by the system. While the convoy is in motion, personal mobility vehicles and autonomous vehicles inside may be moving between convoy vehicles, and therefore, a physical interface will be required to accommodate such mobility between convoy vehicles. In addition, these interfaces will need to provide the ability to attach and detach from the continuous convoy while moving along the a-way. And some inherent electronic communication abilities will be essential between convoy vehicles so various transportation activities can be well coordinated. While the ultimate designs of convoy vehicles may vary, these basic components will likely be standardized.

Convoy vehicles will come in many different sizes, based on the loads they will be carrying. For example, consider convoy vehicles large enough to transport personal mobility vehicles and various containers. In an effort to best utilize the space within these convoy vehicles, nesting techniques will certainly be utilized by placing smaller containers within progressively larger ones, finally containers comparable in size to a person on a personal mobility vehicle. However, convoy vehicles can also utilize "stacking" techniques to facilitate both carrying different size containers and en-route sequencing activities during transit. This can further facilitate optimal use of space within these convoy vehicles.

What exactly is stacking? Stacking refers to the ability of convoy vehicles to have different size containers at different levels. For example, let's imagine we designed our convoy vehicle with a lower level where smaller containers, for example food containers, could be placed. These smaller containers enter the convoy vehicle at this lower level at an a-way access point, and then, while in transit, they move to different convoy vehicles that align with their final destination. But all the while, they only move around this lower level of the convoy vehicle.

Above this lower level of smaller containers would then be another level where personal mobility vehicles and large containers might travel. Inside the larger containers would be nested smaller containers sharing at least part of their trip. And like personal mobility vehicles, these larger containers would also move to the appropriate convoy vehicle using en-route sequencing strategies. Without smaller containers in the way on this upper level, these larger transport items could move within and between convoy vehicles more easily. This is how stacking techniques could be used within convoy vehicles to facilitate optimal space utilization and en-route sequencing activities. You can also imagine another level sized for medication size containers.

In terms of stacking and the use of different layers, each layer within a convoy vehicle could actually be an entire container itself. In fact, several layers might exist within a convoy vehicle accommodating specific needs based on size of vehicles as well as other features. For example, one container layer might provide refrigeration while another accommodates very small autonomous vehicles. Each container layer would then have a standardized interface that connects with the other container layers within the convoy vehicle. These interfaces could thus allow the different layers to be stacked on one another to share power, communications and other services. By having such a design, different layers could be present within the convoy vehicle based on a particular demand at the time. Once again, this greatly enhances efficiency.

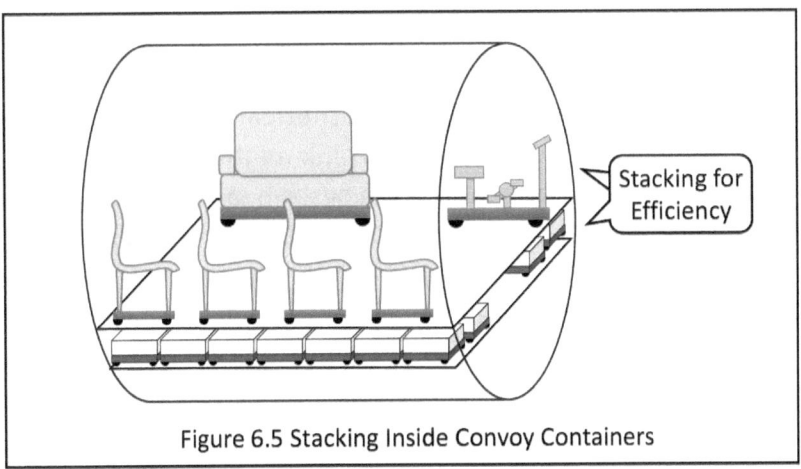

Figure 6.5 Stacking Inside Convoy Containers

En-route sequencing, nesting, and stacking are strategies that not only allow continuous convoys to be time efficient in transporting containers and individuals, but they also facilitate efficient use of space. It should be noted, however, that convoy vehicles by nature must be autonomous vehicles since, at times, they will be operating independently from the rest of the convoy (when detached). Likewise, as noted, standardized interfaces will be needed in order for various activities among convoy vehicles to occur. In addition to providing the means to mechanically attach and detach convoy vehicles from one another, convoy vehicles and their mobility platforms would provide necessary guidance and timing mechanisms. These aspects will be covered in more detail in the chapter dealing with navigation systems.

Traveling in a Continuous Convoy

At this point, you hopefully have a good understanding of continuous convoys and the purpose of en-route sequencing. Both work together in tandem to significantly enhance transportation efficiency by eliminating stops and delays while markedly increasing travel speeds. But how will the system actually work? How do you or any container move from one speed lane to another? What will an a-way station look like? Will such a system be safe given all the activity going on?

In order to address these questions, let's walk through a typical transportation scenario during a hypothetical commute to work. In doing so, we can highlight what we think might be designed within this new autonomous transportation system, and at the same time, we can also consider some alternative options in design as well. As noted in describing autonomous vehicle designs, the final design of components of the overall a-way system will likely vary in different situations. Regardless, we can still present a model that utilizes current technologies and takes advantage of the benefits such a system offers. As a result, this will allow you to appreciate how our proposed model markedly improves upon our existing transportation systems.

Imagine it's a Monday morning, and you are ready to leave your home for work. You collect your briefcase and smartphone, and you load into your personal mobility vehicle that you scheduled for pickup at 7:15 am. Interestingly, your personal mobility vehicle might simply be a mobility platform that collects you while you sit in your transportation chair, or it could involve a mobility platform coupled with an entire container perfectly sized for you and your needs. Such a container could even accommodate your entire family for family vacations. Or, since safety and communication features would be an inherent part of these containers, a transportation container could transport your children safely to a destination, even in your absence, while they remained in continuous contact with you by video and audio, and in any case their container and mobility platform can be tracked for you.

Once in your personal mobility vehicle, you are whisked away from your home through your neighborhood. You may greet neighbors talking, walking, jogging, and riding their own personal mobility vehicles. Or you may hardly notice since while you are traveling you are reviewing a presentation you will be giving later that morning at work. Within a short time, a door slides open and you enter along with other neighbors. The doors close and you feel yourself descending. Soon the doors slide open and your personal mobility vehicle takes you into an a-way station. Depending on where your home is located, this may be a 30-mph lane or a faster lane.

Within a short time, you arrive at the next a-way station. Depending on where your home is located, this may be an a-way station for connecting to a 65-mph lane only, or it may allow connections to 130-mph and 260-mph a-way lanes as well. This is simply a matter of location since a-way station frequency varies for each speed lane. Specifically, a-way stations exist every 4 miles for 65-mph lane access while stations connecting to 130-mph lanes and 260-mph lanes are every 16 and 64 miles respectively. Thus, at some stations, you may only be able to connect to a 65-mph lane initially, while at others, faster lane connections are possible.

Let's assume you arrive first at an a-way station that offers connection to a 30-mph lane only. As you arrive at the station, your personal mobility vehicle automatically moves you onto a large convoy vehicle along with other personal mobility vehicles and autonomous vehicles carrying containers. While you are loaded on an upper level of the convoy vehicle along with larger containers, smaller containers are being loaded on a lower level of the convoy vehicle beneath you using a stacking strategy. Once the convoy vehicle is loaded, it then begins to accelerate as a continuous convoy approaches the station. Note that at full capacity, convoys arrive every three minutes. Just as the continuous convoy catches up, your convoy vehicle matches its 30-mph speed and attaches to the front of the convoy.

Now that you are part of the continuous convoy, your personal mobility vehicle uses en-route sequencing to move to the correct convoy vehicle that correlates with your a-way station exit. Let's assume that your office is several miles away, so your best option is actually to connect to the 130-mph speed lane to minimize your travel time. With this in mind, your personal mobility vehicle has identified that the next a-way station that permits connection to the 130-mph lane is two exits ahead. While you continue to review your presentation, your vehicle automatically relocates you to the correct convoy vehicle that correlates with that a-way station exit. By the time your continuous convoy reaches that station, you are on the convoy vehicle at the end of the convoy. And just before the entire convoy passes this station, your specific convoy vehicle detaches, decelerates, and stops at the a-way station you wanted.

Once you arrive at this a-way station, some containers and passengers on your same convoy vehicle may want to exit while others will want to connect to the 130-mph lane like yourself. Those who wish to exit will do so while you will likely stay on the same large convoy vehicle in your personal mobility vehicle. Additional containers and personal mobility vehicles will also be loading into the large convoy vehicle as they arrive at this station. Once the convoy vehicle is loaded, it then merges onto the 130-mph lane, accelerates, and attaches to the front of a continuous convoy traveling at 130-mph.

This same process would occur for the 260-mph lane as well, and at any time you wanted to exit the a-way, you could simply do so or access different speed lanes on the a-way, or even connect to a different direction. It should be noted that some a-way stations will have options to enter different speed lanes, and as a result, the initial convoy vehicle in which you are loaded will likely correspond to the speed lane you desire. In fact, the system could be designed so that some convoy vehicles never stop as they transfer from the slower lanes to the faster lanes for long distance travelers, while other convoy vehicles provide only travel in specific speed lanes. This is where nuances of design may vary given specific transportation needs.

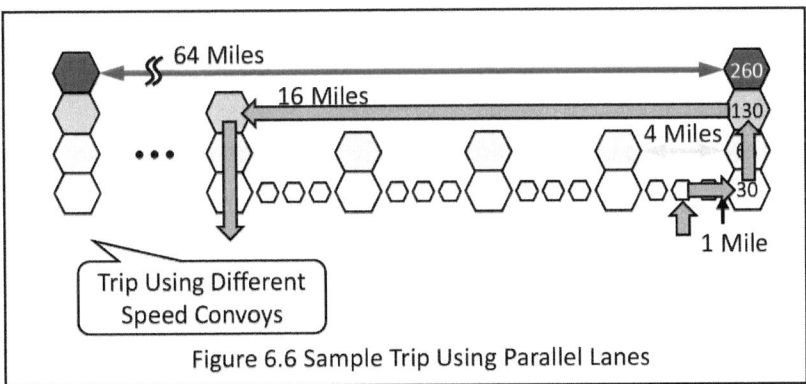

Figure 6.6 Sample Trip Using Parallel Lanes

It should also be noted that passengers as well as other items will likely be located in containers allowing only the containers themselves to be moved from one lane to the next instead of the entire vehicle. For one, different speed lanes may use different mobility platforms, so moving the entire vehicle would not be feasible. Likewise, moving containers would allow greater efficiency of the entire process while mobility platforms remained in place. Also note that you never have access to the actual path the mobility platforms use, the way you are next to the tracks on a subway or train, assuring a safe and quiet ride. The next chapter expands on this with additional features.

In our example, your personal mobility vehicle performed all the necessary navigation to get you to the correct convoy vehicle and continuous convoy, but at the same time, you could have chosen to navigate your transportation yourself once you arrived at the a-way station. Naturally, we will not always know in advance our destination…sometimes, we may just want to explore and roam. In these instances, you can select which convoy vehicle you want to enter on your arrival at the a-way station. In contrast, however, other autonomous vehicles carrying containers will have programmed destinations, which will allow autonomous movement of these containers through the convoys. This feature is one of the ways convoy scheduling can be optimized since the volume of containers requiring transportation in any specific direction will be known in advance.

Practical Advantages of Continuous Convoys

As evident from our previous example, continuous convoys with en-route sequencing provide significantly better efficiency and capacity when compared to our current transportation systems. But the question is how much better? With an a-way that offers continuous entry from any location, access is definitely improved. With lanes traveling different speeds, time of travel is certainly enhanced. And with the use of nesting and stacking techniques, transportation capacity is better as well. While this is intuitive, it is still helpful to quantify the degree by which all of these areas are improved with our new transportation system in order to put things in perspective.

For this purpose, let's compare different modes of transportation involving a trip from Manhattan to Albany, New York. The distance between the two cities is roughly 150 miles, and your options of travel could include a train, a car, or a continuous convoy. For the sake of comparison, we will assume that each mode of transportation travels at 65 mph and that traffic is not encountered for the cars. We will also assume that the train runs once an hour with 10 stops along the way, and that one 10-minute break is required for individuals traveling by car. In contrast, the continuous convoy also serves 10 stations, but does not have any stops or delays by nature of its continuous process.

Given these assumptions, the time of travel for the continuous convoy is 2.3 hours (or 140 minutes) at 65 mph because there are no interruptions in speed during the entire trip. In contrast, the train will make 10 stops resulting in substantial delays. If we allow 1.5 minutes for each stop and include over a minute to accelerate or decelerate at each stop, then the time required for the train to make the trip is 37 minutes longer than our continuous convoy even when traveling the same speed. Lastly, the car will be 10 minutes slower than the continuous convoy simply because of the 10-minute rest stop. This would be an ideal situation, of course, because in all probability, the car would encounter traffic delays, stop lights, and other factors that would prevent it from actually traveling at a continuous speed of 65 mph.

The time efficiencies associated with continuous convoys persist at higher speeds as well when compared to train and car travel. In fact, they are likely more substantial at higher speeds simply because of other factors. For example, cars traveling at 130 mph or 260 mph would not only be unsafe, but managing this in bad weather and with current road configurations would be impossible (not to mention the time delays associated with the car crashes that would occur!). Similarly, for trains, the time required for deceleration and acceleration around each station stop would further hinder its time efficiency when compared to a continuous convoy. Thus, it becomes readily apparent that the time efficiencies associated with continuous convoys increases even more at faster speeds.

How about comparing the passenger-carrying capacity of these three modes of transportation? The capacity of a "superliner" train is 960 passengers. For our trip from Manhattan to Albany, this would mean the train could serve 1,200 people each hour in a trip if we assume that 5 percent of passengers get on and off the train at each stop. If we assume each car carries 2 people including the driver, and we assume 2,000 cars each hour for each lane of a three-lane highway, then cars could serve 12,000 people each hour. In contrast, a continuous convoy on our a-way could serve as many as 60,000 people each hour using a single lane. In other words, the continuous convoy would carry 5 times the capacity of cars on an ideal highway with three times as many lanes, or 15 times the capacity of a single car lane, and 50 times as many people when compared to a superliner train running once an hour.

The efficiencies enjoyed with faster travel times and larger passenger capacities support the use of continuous convoys over current modes of transportation. Factor in the additional personal time enjoyed by passengers as a result of autonomous controls, and this is a no-brainer. But we can also consider the impressive efficiencies continuous convoys provide in energy use as well. We have already discussed the significant advantages in energy efficiency that autonomous vehicles have over cars and trains based on differences in engines, drivetrains, and other equipment. But continuous convoys have additional benefits in energy efficiency as well.

Even ignoring the efficiency advantages over engines and drivetrains, the energy required just to overcome air resistance and wheel friction is a significant factor to consider, especially at higher speeds. Assuming cars have the same air "drag" coefficient as the most efficient production car (the Tesla model S), and the continuous convoy vehicles have the same characteristics as the superliner train (drag coefficient 7.5 times more than the Tesla) the continuous convoy has the advantage of much higher capacity. Also, the energy required per passenger is markedly less for continuous convoys because the air drag is spread over so many people. Specifically, the energy required to move a passenger at 65 mph on our continuous convoy is 24 times less than a car and 6 times less than the superliner train.

This energy advantage can be increased further by detailed design of the convoy vehicles and a-ways. For example, a local convoy between Manhattan and Albany, with 25% of passengers exiting and entering at each station, has 80% greater capacity, and 15% lower power per passenger. This local convoy could travel 180 mph on the same power per passenger as cars at 65 mph. In fact, the power required to overcome air drag and wheel friction for a single passenger traveling 65 mph on a continuous convoy is about 150 watts...less than the lights in a single room of your home!

You can see how the advantages of continuous convoys quickly add up. Energy savings are also enjoyed by not having an entire convoy stopping and starting at each passenger stop. When subways and trains make such stops, additional deceleration and acceleration energies are required, and similarly, energy is wasted while these vehicles are sitting idle at the station without actual transportation occurring. These wasteful inefficiencies of energy use are markedly reduced with continuous convoys. And with convoy vehicles shared so efficiently, the cost and energy required for maintenance and repair is also less. In addition to the benefits that autonomous vehicles offer in energy use and efficiency, continuous convoys further provide many practical advantages in these areas as well.

When considering heavily congested areas like Washington D.C., Boston or Manhattan, you might wonder where these a-ways will be located. One option is suspended above streets or attached to the sides of buildings. With higher speeds, a-ways will naturally require fairly straight tracks to facilitate the movement required during en route sequencing. With this in mind, Elon Musk's Boring Company offers an ideal solution by providing the means by which these can be constructed underground. Just like Musk's Hyperloop, a-ways could be constructed in a manner that avoids interfering with existing properties while facilitating high-speed transportation.

It should also be noted that a-ways provide an excellent system to feed Musk's Hyperloop system. Understanding the Hyperloop could travel at speeds of 750 mph, having frequent stops to load into the Hyperloop would not be feasible. After all, at that speed, it would require over 7 miles for acceleration and deceleration at each stop. Thus, it would make more sense for the distance between Hyperloop stations to be 100 miles or more. However, with a-ways providing continuous access to progressively higher speed convoys, these could provide access to Hyperloop stations for any traveler regardless of where they start their trip.

Imagine traveling on a Hyperloop vehicle from Manhattan to Albany. At 750 mph, the trip would take only 15 minutes! Of course, at this speed, moving around inside the vehicle would be difficult, and en route sequencing may be avoided. However, because Hyperloop stations would be significantly farther apart when compared to lower speed a-way stations, passengers and autonomous vehicles could simply enter the Hyperloop vehicle (at the Hyperloop station) that correlated with their final destination from the start. Combining a-way systems and their continuous convoys with a Hyperloop system would not only be possible but highly favorable in achieving an incredibly efficient and effective transportation system.

Vehicle	Speed, mph	Station Spacing, miles	Capacity, people / hour	Average Power, watts / person
Convoy	30	1	64,000	60
	65	4	65,000	160
	130	16	72,000	400
	260	64	76,000	1,200
Tesla Model S	65	none	4,000	3,500

Figure 6.7 Comparing Speed, Capacity and Average Power

The next time you decide to take the subway or a commuter rail, track the amount of time you could be saving if you could enjoy a-ways and continuous convoys now. As soon as you leave your home, start adding up the minutes. On your way to the rail station, how many times was your car stopped or delayed? Once you arrived there, how long did it take to park and walk into the terminal? Once you had your ticket, how long before your train arrived? Once on board, how many stops were made, and how long did each stop last? And once you arrived at the subway station you wanted, how many other forms of transportation were needed to get to your final destination?

With each of these considerations, you will see specific areas where transportation efficiency can be improved. Continuous convoys, in combination with a-ways and autonomous vehicles, offer the means to address these problems with today's existing technologies. Not only is access markedly improved along with the elimination of many unnecessary delays, but the speed of travel and capacity of occupancy is significantly better as well. And all the while, less energy is required, less pollution produced, and better use of our time is enjoyed.

7
Autonomous Elevators

DO YOU KNOW the name Roald Dahl? Perhaps not. But he was the author of the fantasy novel *Charlie and the Chocolate Factory*.[66] One of the most pivotal scenes, and in fact the climax of the book, involved the protagonist of the book, Charlie, riding in the "Great Glass Wonkavator." Unlike normal elevators that only traveled up and down, the Wonkavator could travel in any direction…sideways, slantways, long ways, backways, square ways, front ways and any other way you can think of!" In fact, the concept was so powerful, Dahl wrote a sequel to the book entitled *Charlie and the Great Glass Elevator* where the characters traveled into outer space and encountered Martians.[67] Clearly, Mr. Dahl was a visionary when it came to elevator travel.

For decades now, societies have used elevators to travel up and down large buildings. Despite the many limitations and inefficiencies that exist, we continue to rely on traditional elevator designs to move us from one floor to another. Like today's automobiles, however, these designs utilize tremendous amounts of energy while

---

66 Dahl, Roald. "Charlie and the Chocolate Factory. RoaldDahl.com, 2015. Retrieved from http://www.roalddahl.com/roald-dahl/stories/a-e/charlie-and-the-chocolate-factory

67 Ibid.

consuming an abundance of space, and all the while, they are associated with delays and frustrations. These designs even constrain architectural pursuits since buildings can only reach as high as an elevator will allow its occupants to travel. Don't you think it's time for a better idea when it comes to elevators? How cool would it be to actually have a Wonkavator to get us where we were going!

In this chapter, we will present our concept of an innovative elevator design. Not only can these elevators move in a variety of directions, but they are also autonomous, safer, and markedly more efficient in comparison to elevator designs currently. Such autonomous elevators enable transportation systems to be significantly improved, as you will see, and the ripple effects this can have on society are worth noting. What might have simply been fantasy in the world of Roald Dahl can actually be reality given today's technologies. Incorporating autonomous elevators into our overall autonomous transportation model offers a logical extension of a system given the limitations and frustrations each of us experience with the elevators of today.

The Problems with Today's Elevators
You have just arrived in the hotel lobby of a 48-story, high-rise building after negotiating planes, trains, and automobiles on yet another one of those mandatory business trips. After registering and obtaining your room key, you load yourself once again with your shoulder bag and suitcase and make your way to the bank of elevators…the last leg of a very long journey. You instinctively press the "up" arrow on the first set of elevator buttons you see and then not-so-patiently wait for the elevator to arrive. Only after several seconds do you then realize the button you pushed only calls for the elevator traveling to floors 2 through 24. Unfortunately, your room is on the 30th floor, and you have to find the other set of buttons for those elevators.

Eventually, you hear the "ding" announcing that the correct elevator has arrived. After a few more wasted seconds, you find the one with the open door, and you squeeze you and your luggage in the midst of several other hotel guests. Of course, none of these individuals are traveling to the same floor that you are, so you again bide your time as the elevator door opens and closes at each of the lower floors beckoned. Some guests exit, others enter…and all the while, you dream of sprawling out on the bed of your hotel room for some much-needed peace. What could have been an expedited end to your long day of travel has turned into another undesirable transportation experience.

The scenario depicted is certainly not the most common experience with elevators today, but some of the frustrations described are quite common. Waiting for an elevator in a crowded building challenges even the most patient person, and these delays have notable implications during emergencies where buildings must be quickly evacuated. Transit delays are just one of the problems associated with elevators related to poor time efficiency. Time is required for elevators to load and unload, and even the opening and closing of elevator doors consumes some degree of time. If one of the goals of elevators is to get us to and from different floors of a building quickly, tremendous room for improvement exists.

Time delays are not the only major problems related to modern elevators. Have you ever thought about how much real estate elevators occupy within a building? In tall buildings, as much as 20 percent of the total floor space can be consumed by elevators and their shafts. If we are talking about such a building in the heart of New York or Chicago, you can appreciate the expense associated with this loss of space. At the same time, architects can only design buildings a certain height since they must accommodate the constraints of today's elevator systems. Because elevators rely on a cable pulley system with counter weights, building heights are limited by the strength of these cables. Not only are elevators currently poor in time efficiency, but they are also space inefficient as well.

In our previous discussion concerning automobiles, we highlighted how much energy is wasted with the current models of cars, engines, and roadways. One of the major factors in this regard pertained to the weight of our standard vehicle. Elevators suffer from the same problem. First of all, elevators are quite heavy, and this is particularly true for freight elevators that have to transport larger materials up and down a building. In addition, elevators must move this weight against gravity as opposed to cars, which move weight horizontally. Because of these factors, elevators alone can account for 40 percent of an entire building's energy consumption when they are being used at peak times. That is a pretty sizable figure considering a facility's air conditioning, lighting and electronic requirements. Here again, significant room for improvement exists.

Lastly, we come to some of the more functional issues related to modern elevators. For those of you who live in a high-rise apartment building or condominium complex, you can appreciate that your furniture did not likely arrive to your dwelling by way of the tenant elevators. Instead, larger (and less aesthetically pleasing) freight elevators had to be used. While this reflects a practical need in many ways, it also requires two different elevators to be designed for these buildings. Similarly, you may have also experienced a broken elevator that requires you to take the stairs instead (we are naturally reminded of the perpetually broken elevator on the television program *The Big Bang Theory*). Not only do current elevator systems require complete shut-downs for repairs, they also require ongoing maintenance that may similarly interfere with their use.

As you can see, a number of problems are present when we consider modern elevator systems. Most of the time, we simply overlook these issues and tolerate the inconveniences. But better options and solutions exist. Much of the time, space and energy issues can be resolved through alternative elevator designs, and functional improvements can be similarly addressed in the process. In fact, alternative elevator systems could even connect with other transportation systems to create a completely new experience for us all.

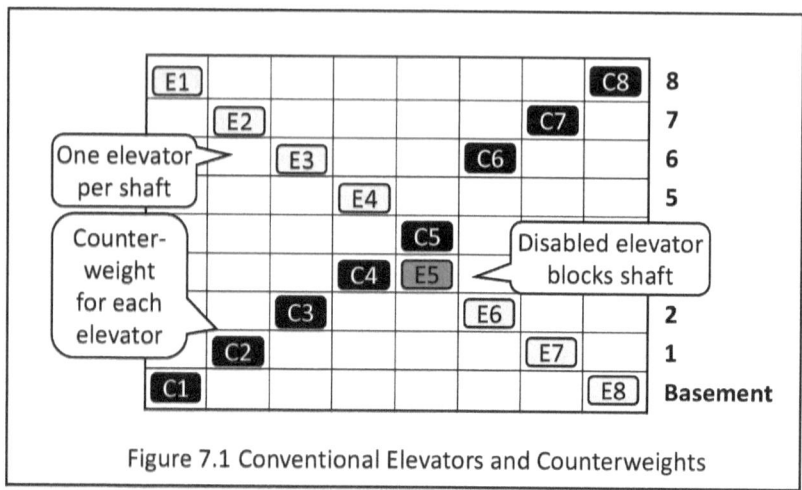

Figure 7.1 Conventional Elevators and Counterweights

Introducing Autonomous Elevators

Having identified the problems associated with today's elevator systems, can't we design a new system that either eliminates or greatly reduces these issues? In fact, wouldn't it be ideal if we could simply apply the designs of a-ways and autonomous vehicles to an elevator concept? Guess what…we can! All we have to do is turn our a-ways on their end…by envisioning these structures vertically instead of horizontally, we can achieve the same efficiencies we described earlier with a-ways. And in the process, we can greatly improve how we and other items move around within buildings and high-rise structures.

Of course, simply turning a-ways on their end would not magically create autonomous elevators. However, this provides you with an opportunity to envision how autonomous elevators might exist. The first issue that would need to be addressed relates to gravity. How can autonomous elevators travel up and down like autonomous vehicles without dropping straight to the ground? If you recall, autonomous vehicles have wheels located on one side (the bottom) with which they contact a-way surfaces. Of course, a-ways only have one riding surface. If we think about autonomous elevators, however, the lanes in which they travel could have more than one riding surface. With this in mind, we could place wheels on opposite sides of an autonomous elevator to create opposing forces of pressure to prevent slippage that would occur due to gravity.

While this concept may sound a bit foreign to you, it is the same technique rock climbers have used for centuries. When faced with a narrow crevice through which they must ascend (typically called a rock chimney), climbers will use various parts of their bodies to press against both sides of the crevice to prevent them from slipping downward. One technique is called "stemming" where each leg is on the opposing faces of the rock chimney pressing outward to maintain position. Another is called "frogging" where the back and feet press against one rock face while the hands and knees press on the other.[68] In both cases, this technique allows stability against gravity so that movement upward may continue.

By having such a design for autonomous elevators, movement can be provided using the same electrical systems used in transporting autonomous vehicles along a-ways. Pulleys, cables and counterweights would no longer be needed. Likewise, when autonomous elevators are descending, gravity would provide the power for motion while energy from regenerative braking during deceleration could be recaptured for upward movement of other autonomous elevators at the same time. This type of design would be incredibly more energy efficient than current elevators, and the height constraints of buildings present with today's elevator designs would be eliminated.

Of course, the advantages of autonomous elevators in terms of energy and space efficiency do not stop there. Even at the busiest times, a conventional elevator is moving less than half the time it is

---

68 Ellison, Julie. "Learn this: Conquer chimneys." Climbing.com, 2014. Retrieved from https://www.climbing.com/skills/learn-this-conquer-chimneys-with-tips-from-rob-pizem/

in operation, and conventional elevators only have one elevator per shaft. In contrast, autonomous elevator designs can have many vehicles per lane with each providing service at the same time. With this in mind, autonomous elevators use much less than half of the floor space of conventional elevators and significantly less energy while delivering passengers and objects faster. From energy and space efficiencies alone, autonomous elevators offer significant benefits over current elevator designs.

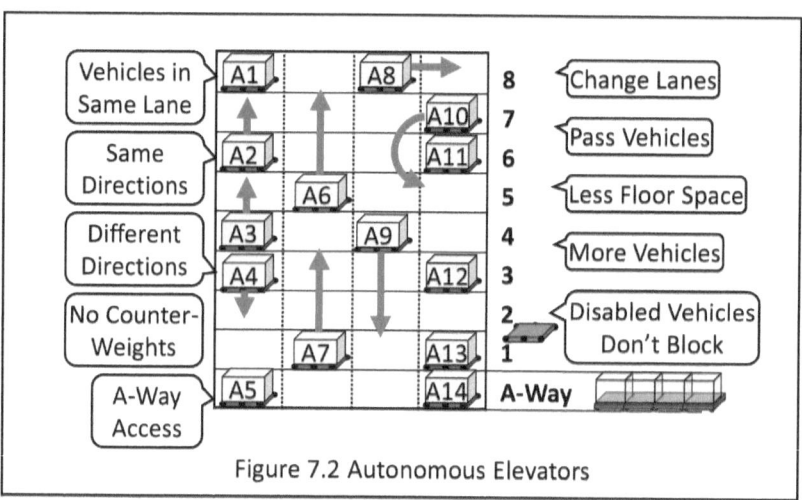

Figure 7.2 Autonomous Elevators

One concern some people may have regarding this proposed elevator design is safety. However, our design also incorporates features that greatly enhance the safety of elevator use in the event of power failures. Some of you may recall the "great blackout" in New York City in 1977 when roughly 7 million people lost electrical power for more than a day. Thousands of people were trapped inside elevators during this time causing a great deal of fear and panic.[69] This again occurred in the New York area in 2003 as well as most of the

---

[69] Martin, Phillip. "When the lights went out in New York City: A tale of two nights." PRI.org, 2015. Retrieved from https://www.pri.org/stories/2015-07-18/when-lights-went-out-new-york-city-tale-two-nights

Northeast causing many to be entombed within elevators for hours a second time.[70] Given this history, you can see why some people may worry about elevator safety.

Our solution is to have a fail-safe mechanism in place to prevent cabin descent or entrapment. This is actually a "fail-functional" feature since our design allows the elevator to still function during electrical failure so that elevator occupants can exit the compartments safely. Having explained the mechanism that uses opposing pressures to prevent elevator slippage, this same technique would be used when electrical or mechanical failures occur. Within the working design, an increased amount of pressure would be imposed on both sides of the autonomous elevator lane by the elevator itself as soon as electrical power was lost. This would "lock" the autonomous elevator in place securing its position as well as the safety of any passengers or contents. At that point, unless power was immediately resumed, the pressure would then be slowly reduced to allow a gradual, controlled descent of the autonomous elevator allowing a safe exit.

As you can appreciate, autonomous elevators would offer major advantages in energy efficiency, safety, and building design. But another major benefit would also be gained...the ability to move in any direction you wanted! Since our autonomous elevator has wheels on opposing sides and electricity-supplied mobility, it can now move up, down, left, right and diagonally. This would permit you to not only move among different floors of the building but to different areas of the same floor as well. This feature would also permit several autonomous elevators to occupy a single elevator lane (when more than one lane existed). If one autonomous elevator stopped, the other one could simply bypass it by switching lanes, no different than a vehicle passes another one on the road.

---

70 Siegel, Joel, and Corky Siemaszko. "Blackout hits New York City and Northeast in 2003." New York Daily News, 2015. Retrieved from http://www.nydailynews.com/news/national/blackout-hits-northeast-united-states-2003-article-1.2322074

With this ability, we can eliminate many of those frustrating time delays currently experienced when we move up and down the floors of a building. Likewise, it opens up all sorts of possibilities in relation to elevator lane designs. Some lanes might be dedicated for descent while others can be dedicated for ascent. This might even vary at any given time depending on the volume of passengers and packages moving in a specific direction within a building. For example, during an emergency evacuation of a building, all but one of the autonomous elevator lanes could be committed for people evacuating leaving the other for autonomous elevators to go back up to get more people. Also, that lane can be for emergency personnel to access the building. Here again, this would offer enhanced safety through flexible design.

    This previous example involving emergency evacuations highlights the potential energy efficiencies of autonomous elevators once again. Assuming autonomous elevators would be relatively lightweight and energy efficient, the potential energy of people descending on autonomous elevators from higher floors can be captured through regenerative braking. This energy could then be used to send empty autonomous elevators up again to collect additional occupants even in the absence of external electrical power. This is certainly not the case with today's elevator systems.

    The last feature to be described concerning autonomous elevators involves the actual elevators themselves. If you recall, one of the inconveniences of current elevators involves their loading and unloading. While this may be a minor issue when it comes to guests in a hotel lobby, it is a major concern when having to unload a moving truck and using the freight elevator in a high-rise building. What if this step could be eliminated altogether? In other words, what if the container carrying all your furniture could simply be transported to your floor without this intervening step of unloading and loading?

    Just as the container and mobility platform represent two components of our autonomous vehicles, similar features would exist for autonomous elevators. In fact, the same containers arriving via a-ways could simply be transferred onto autonomous elevator mobility platforms that had wheels on their opposing sides. As an autonomous vehicle arrived (presumably at the ground or underground level), the container would be removed from the autonomous vehicle mobility platform and placed on an autonomous elevator mobility platform. As a result, unloading would not have to occur until the actual destination in the building was reached. And by communicating the final destination of the container in advance, the process of moving from a-way to autonomous elevator to the destination floor of the building would be seamless without delays or interruptions.

This type of design has many advantages beyond transportation efficiencies. For one, containers designed for specific contents would be used to meet specific needs. This might include refrigerated containers for food items, padded containers for furniture and appliances, and aesthetically pleasing and air-conditioned containers for passengers. Likewise, passengers or packages would be clustered into common destination containers to speed transportation. At the same time, individuals arriving at an autonomous elevator could enter an awaiting container on site to travel to their destination within the building, without detaining a mobility platform or using a lane during loading.

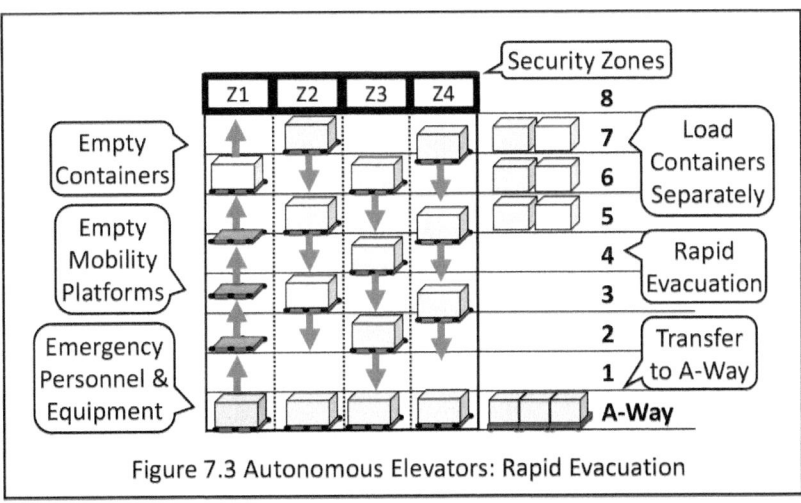

Figure 7.3 Autonomous Elevators: Rapid Evacuation

Autonomous elevators offer many advantages in transportation when compared to current elevator systems. Significant efficiencies can be gained in time, energy, space, and function simply by implementing a new design that applies many of the same concepts utilized in a-ways and autonomous vehicles. But what would such a system actually look like? And how would it work? In the next section, we will demonstrate how autonomous elevators can actually integrate with a-ways and function throughout building complexes not only for passengers but for a variety of items big and small.

Envisioning an Actual Autonomous Elevator System

Understanding the basic features of our autonomous elevators, you might wonder how these structures might be utilized within a multilevel building. How would they integrate with our a-ways and autonomous vehicles as well as with our current mobility needs? For the purpose of demonstrating this in greater detail, let's assume you are moving into a new apartment on the 15th floor of a building in the heart of a downtown urban area. The day has arrived for you to move, and being well prepared, you have made all the necessary arrangements. In addition, you want to arrive ahead of time to meet the actual movers, so you can indicate where you would like your furniture placed.

In your preparations, a moving company has already come the day before and placed all your furniture and other household items in your prior dwelling within their designated reusable containers, in place of the conventional cardboard boxes. For fragile items, padded containers were utilized; for some of your food items, refrigerated containers were selected; and for your furniture, other containers were chosen. These containers are then nested within larger containers and placed on mobility platforms, so they can be transported to your new apartment by way of the a-way. And after programming these mobility platforms with the appropriate data, all of your things are ready to relocate to the new building autonomously.

With this peace of mind, you decide to make your way over to your new apartment as well. After entering the a-way on a personal autonomous vehicle, you are whisked away with the most efficient a-way lanes are used to speed you to your destination. While en route, you text the movers that you are en route, and you also log into a secure scheduling application to track the transportation of your household items. Before you know it, your autonomous vehicle arrives near your new apartment building, and your autonomous vehicle carries you to the ground floor of the building.

On your arrival to the building, you don't have to bother getting up just yet. Because your trip has already been programmed ahead of time, an autonomous elevator mobility platform arrives just in time to meet your autonomous vehicle. As the two interface, the a-way mobility platform transfers the container in which you are traveling to an autonomous elevator mobility platform. Once this is done, the autonomous elevator ascends in one of the elevator lanes to take you to the 15th floor. In the meantime, you continue to check your emails and texts while making plans for a housewarming party. Because the interiors of both the a-ways and autonomous elevator lanes are smooth and enclosed, your transportation throughout has been comfortable allowing you to easily perform these tasks.

Your apartment building is equipped with four autonomous elevator lanes. Two are assigned to ascending while the other two are for descending. In addition, one of the lanes for ascending and descending is for local traffic moving between a couple of floors while the other lanes are express lanes for longer distance routes, similar to adjacent lanes in a-ways for different speeds. Because you are traveling to the 15th floor, your autonomous elevator immediately shifts into the express lane to speed your arrival. And just before reaching the 15th floor, it shifts back into the local lane as it slows for your destination arrival. As a result, you arrive to your apartment in record time.

Since autonomous elevators do not stop when traveling in the express lanes, multiple autonomous elevators can use the same lane at once. Likewise, should an autonomous elevator encounter another autonomous elevator that is stopped at a particular floor in the local lane, it can simply move over to the express lane temporarily to bypass the stopped autonomous elevator. But today, this isn't necessary, and your autonomous elevator travels to your floor without having to maneuver around other autonomous elevators. On your arrival, the container opens, and you travel to your actual apartment on your personal mobility device. The movers arrive moments later, and you begin showing them where the major items need to be placed. And just as you complete the instructions, the containers carrying all your household items begin to arrive.

In terms of your other items, many of these containers used the same set of autonomous elevator lanes that you did in arriving to your apartment. While their containers may have had different features than your container, they were still able to use the standard local and express lanes if their size was reasonable. However, your apartment building also has a set of larger autonomous elevator lanes to accommodate more sizable items like furniture and appliances. Like the autonomous elevator lanes you used, these lanes also have ones dedicated for ascending and descending as well as extra lanes for bypassing stopped autonomous elevators. In fact, this set of larger autonomous elevator lanes has been constructed on the outside of the building, sort of like the platforms used by window washers, allowing preservation of interior space, and able to transport large items such as grand pianos and construction materials.

As you can see, the integration of the autonomous elevator system with the a-way system makes traveling to specific locations in a building much faster, autonomous, and without interruptions or frustrations. There is no waiting to load and unload onto an elevator, no delays while others enter and exit at various floors, and no demands of your time if your destination has already been programed into the system. Likewise, this system inherently accommodates greater security and privacy abilities. Because you and your items remain in the same container from your origination point to your final destination, unwanted intrusions can be avoided while making your journey more secure and safe. And when you are finished, the containers are sent off for reuse.

It should also be noted that the previously described example in using the autonomous elevator design primarily allowed transportation up and down with temporary lateral movement when switching between local and express lanes and when bypassing other autonomous elevators. However, lateral movement as well as diagonal movements are also possible in such designs allowing passengers and items to move to different areas on the same floor safely and privately. For newly constructed buildings, this type of architectural planning could allow a variety of innovative designs. Because traditional elevator shafts are no longer required, autonomous elevator lanes could be placed anywhere within the building. And, as in our example, placement of lanes outside the building can similarly be done, which might be ideal for older structures.

Finally, autonomous elevators have one other major advantage in comparison to today's elevator systems. Suppose all your plans for moving were done flawlessly, but on the day that you decided to move, the freight elevator was broken. What options would you have to get all your furniture up to the 15th floor? None, unfortunately... at least that day. In contrast, if your building had an autonomous elevator system, any malfunctioning unit, which essentially would be just the mobility platform, could be easily removed and the remaining ones continue to provide service. Meanwhile, the broken mobility platform could be repaired outside of the autonomous elevator lanes. As a result, your moving plans could proceed without any unexpected delays or inconveniences.

As you can see, autonomous elevators greatly enhance transportation abilities throughout buildings and other indoor structures while providing notable improvements in various efficiencies. The majority of our common frustrations that we encounter with today's elevators could be eliminated or significantly reduced in the process. While this example of using autonomous elevators to help you move into a high-rise apartment building highlights many of these advantages, autonomous elevators improve transit in many other situations as well. By simply applying many features that exist in a-ways to autonomous elevators, we can further enhance the transportation systems of the future.

At first glance, many of the concepts presented in this chapter may seem beyond our technological abilities, but this could not be farther from the truth. In fact, other companies have already considered expanding the movement options of elevators. ThyssenKrupp recently introduced its "Multi" elevator that is able to move up, down, sideways, and slantways. Instead of using cables and pulleys, the Multi uses magnetic levitation to guide the elevator along its path similar to the mechanisms described in Elon Musk's Hyperloop system. By placing strong magnets on the elevator container, a magnetic coil situated along guard rails is used to direct and move the elevator to its destination in any direction.[71]

While the Multi is certainly a big step in the right direction, autonomous elevators offer additional benefits that this design cannot. For one, using the same technologies as the rest of the transportation system significantly reduces costs and increases maintainability. Also, autonomous elevators could easily interface with a-ways and autonomous vehicles streamlining transit not only throughout a building but between locations as well. Likewise, autonomous elevators avoid unnecessary unloading and reloading that would still exist with the ThyssenKrupp model. And in many cases, passengers and items could remain within their own transportation containers until they arrived at their final destination with autonomous elevators. In

---

71 Stinson, Elizabeth. "The sideways elevator of the future is here, and it's wild." Wired.com, 2017. Retrieved from https://www.wired.com/story/the-sideways-elevator-of-the-future-is-here?mbid=nl_7917_p1&CNDID=22832787

addition to convenience, this offers many opportunities for enhanced security and privacy protections. Because of these features, the autonomous elevator provides tremendous advantages for advancing transportation.

8
Linear Cities

WHAT EXACTLY IS a city? One definition describes a city as a large concentration of people with a complex infrastructure to support them and their needs. From food to supplies to housing, a city provides an array of goods and services to its residents and businesses, but none are more important than transportation. After all, without transportation, you can't get your groceries, furniture, or appliances. And without transportation, it's pretty difficult to get to work. Effective transportation systems are essential to any city.

This seems pretty obvious, of course, but what may not be obvious is how transportation systems influence urban designs. Think about most cities today. Neighborhoods and urban districts are often situated where transportation is most abundant. In some cases, transportation options developed in response to popular attractions in these areas, but in other instances, the convenience of transportation made such areas more desirable. In addition, buildings and offices are required to meet transportation needs. Parking lots, parking garages, and frontage access are all influenced by modern transportation systems. Even the urban footprint is affected by the type of transportations systems in place.

In this chapter, we will take a look at the type of urban designs and developments that can occur with the introduction of a-ways and autonomous vehicles. From opportunities related to a-way structures and infrastructures, to changes based on its inherent efficiencies in travel and transport, new urban designs can be expected to emerge. And while the exact design will likely vary greatly based on population, regional and individual preferences, some predictions can be made based on obvious incentives as well as historical trends. These predictions will be provided as a means to highlight the significant role transportation plays in future urban design.

Looking to the Future by Looking Back

If you look around the world today, you will notice that the vast majority of the major cities and metropolitan areas are situated along ocean, rivers, lakes, and bays. New York, Chicago, London, Athens, Hong Kong and San Francisco are some that quickly come to mind. It seems rather obvious that these sites were selected because they allowed easy access to water transportation. Cargo could be readily shipped to and from these cities, and as a result they flourished. And as they flourished, jobs became more abundant, and increasing numbers of people settled in the area.

Of course, waterways were not the only form of transportation that influenced city development. Railroads allowed cities to develop along railway stations (sometimes remote from waterways and oceans) where local commerce was needed. Likewise, railroads expanded the urban sprawl of major cities, especially as commuter rails allowed people to travel more quickly to their jobs from the suburbs. The automobile and highway systems also helped cities expand. Like railroads, small towns developed along highway interchanges, and increasing numbers of people could now live more affordably outside the city. Even corporations followed suit in some cases since land was less expensive, and talented workers now lived outside city centers.

Why is this important? Because transportation is a major factor in determining the location, shape, and size of a city. In fact, there is a general rule of thumb that helps you predict how large an urban area may be. Historically, cities tend to expand until the average commuting time is one-half hour per day. When travel was by foot, towns were much smaller in size. When horses offered faster travel, cities became larger. And as automobiles and trains were introduced, major urban sprawls emerged. Particularly when considering metropolitan areas, existing transportation systems and technologies have had tremendous impacts on city designs and development.

From what we have just explained, you can appreciate that the increasing efficiency of transportation leads to expansion of major cities. If that is the case, then why are we currently seeing a migration of people back to city centers? For one, the inefficiencies of our current transportation networks serve as a major incentive. Why sit in traffic for an hour when you could walk a few blocks downtown to work? In addition, many people are choosing to return to urban centers because of the concentration of services and networking located there. While people in the suburbs enjoy less expensive housing costs and some services, they do not have easy access to the rich environments of city centers.

Unfortunately, living downtown in most cities is not inexpensive. The wealthy can better afford to make these moves as can single individuals who are willing to live in small studio apartments or share a space with friends. Richard Florida has identified these individuals moving back into urban centers as the "creative class" seeking to interact, collaborate and network with others. This has important implications in fostering innovations and creativity, but our current transportation system offers some obstacles in achieving these socially rich environments. As we will discuss later, a-ways offer many advantages in this regard.

These obstacles involving our current transportation system highlight an inherent, underlying truth. Being in close proximity to transportation systems is certainly desirable. It allows us to get to where we are going faster and with less hassle, most of the time. But proximity has many undesirable features as well. Because of demand for proximity, housing closest to transportation access points is often more expensive. Also, increased noise, pollution, and crowds typically increase the closer you get to transportation hubs and thoroughfares. In fact, my wife will often refuse to stay overnight in the heart of some cities because of the noise. Thus, despite wanting to have fast and easy access to transportation systems, the closer we get, the more negative features we encounter.

The good news is this traditional "proximity tension" involving transportation systems does not have to exist. As previously described, a-ways are enclosed, utilize electrical power for energy, and use multiple travel lanes for vehicles of various speeds. Likewise, the design of these a-ways permits continuous access for people and items alike. As a result, problems related to noise and pollution are essentially eliminated, and opportunities to live near easy transportation access are significantly increased. Because of these features, urban development can enjoy the positive attributes of being close to transportation channels without the typical major "down-sides" that have been traditionally present.

In terms of our general rule of thumb that cities must allow access to key areas within half an hour, this still applies with a-ways as well. But without the negative effects associated with today's transportation system, the design and development of cities can move in completely different directions. Not only can much larger populations be in close proximity to the transportation they need, but they can also enjoy much richer and fulfilling environments to meet their own personal needs. And all the while, costs of living are significantly reduced while travel efficiency and quality of life markedly enhanced.

From the Ground Up

We have a friend who hosts a large art event every year in various cities. The unique aspect of the event is that it is typically held in old industrial warehouses, which creates a very raw and creative feel for the event. During the recent recession, finding these warehouses for temporary lease was relatively easy. But as the economy has recovered, it has become more challenging to find locations for the show. Without thinking too much about it, our friend learned to follow a city's railroad tracks in looking for these locations. After all, even in cities where the railroads are no longer active, the industrial sector is still typically located where rails traveled previously so their goods could be easily transported.

While modes of delivery have changed over many decades, warehouses and industrial manufacturers still want to be in close proximity to transportation systems. Have you noticed the number of large warehouses located near superhighway exits in metropolitan areas? These locations enhance customer service through faster delivery of products while reducing costs of delivery. In other cases, the locations provide better access for customers. Think about IKEA and Costco, for example. Because transportation costs represent significant expenses for most industries, locating the site of production close to transportation systems is important.

But transportation access points are highly desirable. As a result, property values are typically higher near these areas because industries, retail stores, and commuters prefer these locations for obvious reasons. For many industries, one way to keep costs down has been to find locations along superhighways far enough removed from urban areas. In fact, this was initially a strategy employed by some major warehouse suppliers like Amazon. But as the demand for next-day (and even same-day) delivery and easy customer access grew, these companies were forced to find locations closer to city centers, and naturally, their costs grew significantly.

So, what does this have to do with a-ways and autonomous vehicles? Actually, a lot. Like railroads and superhighways, a-ways offer industries the transportation systems they need. But a-ways accomplish this much more efficiently and cheaply. Whether products are being shipped across the country or a few miles away, a-ways provide a number of pathways to get items to their destination at a fraction of the cost of today's transportation models. Naturally, industries will want to be in close proximity to a-ways to enjoy the benefits these pathways provide.

With industries wanting to be close to a-ways, the next decision involves where they should locate their operations. Historically speaking, industries have wanted to be as close to transportation access points as possible to facilitate fast transport of their goods to suppliers or customers. But older transportation models have limited access, which increases the cost of property values where industries want to be. A-ways, in contrast, do not have these limitations. a-way access is essentially continuous, and this greatly increases the number of sites that industries will be able to consider for their locations. This increase in accessibility will mean a greater supply of possible locations that would be desirable, and this will significantly reduce location costs for industries.

Let's take this a step further. Current models require industries to construct factories and large warehouses adjacent to superhighways, roads and railways. In addition to the footprint of land required to accommodate such warehouses, additional space is consumed by trucks, cars and parking lots. But with the use of autonomous vehicles, trucks and cars are no longer required, and likewise, neither are parking lots. More importantly, warehouses would no longer have to be built adjacent to a-ways. In fact, they could actually be built right on top of them!

Think about it for a second. We now have an enclosed a-way with continuous access points along the way. If we could construct industries to be immediately on top of these a-way structures, not only is access to transportation readily available, but the amount of land now required for both transportation and industrial operations is markedly less. And don't forget our a-ways have an abundance of other infrastructures that can be shared by these industries. Electrical power, water supplies, communications and other infrastructures present within a-ways could also supply industries with needed supports while again reducing costs of installation, maintenance and repairs.

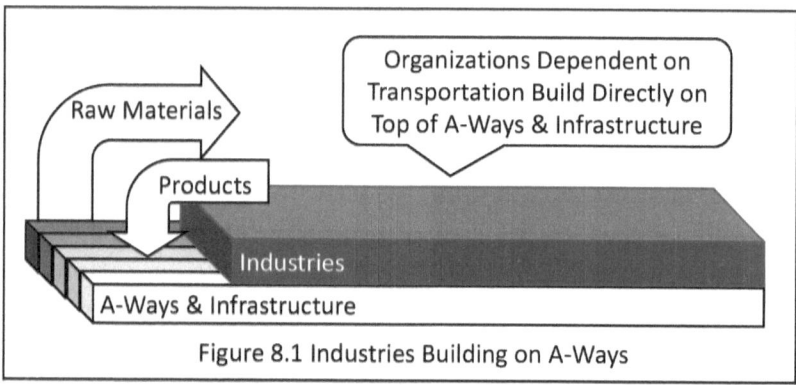

Figure 8.1 Industries Building on A-Ways

Sometimes, thinking of new paradigms can be challenging when all we know is one way of doing something. Today, we are accustomed to having roads and highways provide the majority of our transportation needs. Because of this, it may be hard to imagine having industries and warehouses constructed on top of a transportation system when all we know are large tracts of land consumed by industrial complexes. But from every perspective, pursuing such a model makes perfect sense. A-ways allow such new paradigms to become reality.

Expanding the Paradigm

Since the 19th century, it has been well appreciated that building up rather than out offers some significant benefits, especially in concentrated urban areas. With metropolitan land space at a premium, and with the manufacturing of high-strength steel, architects and builders began constructing skyscrapers and multilevel buildings to reduce the geographic footprint. The same approach was applied to urban parking needs as underground parking levels and parking garages similarly saved land space. And by building vertically instead of horizontally, fewer construction materials were needed.

High-rise buildings offer many advantages. Building "up" is economically better than building "out" in congested city centers (in fact, building out is not even an option in many instances). At the same time, the expansion of offices and apartment buildings vertically save a great deal of time. Residents living downtown no longer have to sit in traffic traveling from suburban areas, and commercial activities were much more efficient since businesses are closer to one another. You might have to wait on an elevator at times, or perhaps pay a premium for parking, but it is better than spending hours on the highway each week.

So, why do skyscrapers and multilevel buildings make sense? Because they allow better use of our time and available space. Saving time and space also means reduced costs in one way or another. As we have seen by relocating industries on top of a-ways, transportation becomes faster since industries are in close proximity to transportation access points. Land space requirements are also eliminated since the same land used for a-ways provides the same footprint for the industrial complex. Shared walls and roof mean more efficient heating, cooling, pipes and wires. And costs fall as construction requirements are fewer and infrastructures are shared.

Now suppose you're a restaurant owner or a small business retailer. You have inventory needs, but you don't want to hold onto a bunch of supplies in advance that you won't need until later. At the same time, you don't want to be stuck waiting for your supplies while customers wait or decide to go elsewhere. This is the bane of inventory management. In a perfect world, you could simply receive your supplies from various manufacturers right when they were needed. Like our small autonomous vehicles carrying medications, supplies would arrive "just in time." No extra holding costs, no wasted space, and no customer service delays.

With this in mind, wouldn't retailers and small businesses also want to be right next to a-ways? Of course! The transportation system of a-ways would offer quick access to inventory supplies in addition to rapid delivery services to customers if needed. As a result, places of business could also be located above a-ways as previously described for industries and manufacturers. In fact, restaurants and retailers could be built vertically above key industries that provided them with necessary supplies, and they too could share a-way infrastructures to lower their land, construction and operating costs.

Naturally, both industries and small businesses will need employees, right? If industries and stores are situated above our new autonomous transportation system, then it only makes sense that people would want to be close to these areas to minimize their commutes to work. Certainly, the expanded accessibility of a-ways facilitates shorter travel times to the job, but at the same time, residential areas can improve on this even further by being located above the industries and businesses that supply individuals with jobs. By expanding communities vertically, land space is again preserved, infrastructures again shared, and access to jobs and city services readily available.

Do you think this sounds far-fetched? If you have recently walked downtown in most city centers, you have likely noticed an increasing number of mixed-use spaces. On the ground level of a high-rise building may be clothing stores and coffee shops with an athletic gym above on the second level. Above these levels are either apartments, condominiums or office suites. Why? For many of the same reasons. Ground level stores offer better access to customers as well as suppliers along city streets while homes, offices, and other services above get to share space and infrastructure costs. Better access, lower costs, and less space required drive these trends. A-ways just take this further by adding a much more advanced transportation system to the mix.

Just as industries and businesses would attract housing through jobs offered and services rendered, other community services would likely follow these trends. Hospitals, recreational activities, municipal governments, schools, and entertainment venues would benefit from being close to residential neighborhoods, businesses, and industries. These community services would help attract residents, and they would enjoy reciprocal relationships with other businesses and industries. All the while, land space is preserved, infrastructure construction reduced, and transportation access and efficiency enhanced.

Unlike current transportation systems, a-ways have complementary structures that facilitate vertical growth of industries, services, neighborhoods and communities. Autonomous elevators, as previously described, permit rapid and autonomous mobility up and down the different levels of these communities in addition to convenient access to a-ways. As a result, traveling in vertical or horizontal directions is easy and efficient for items and individuals alike. This allows families and businesses to freely choose the location that best meets their needs while still enjoying great transportation, reasonable expenses, and an array of services.

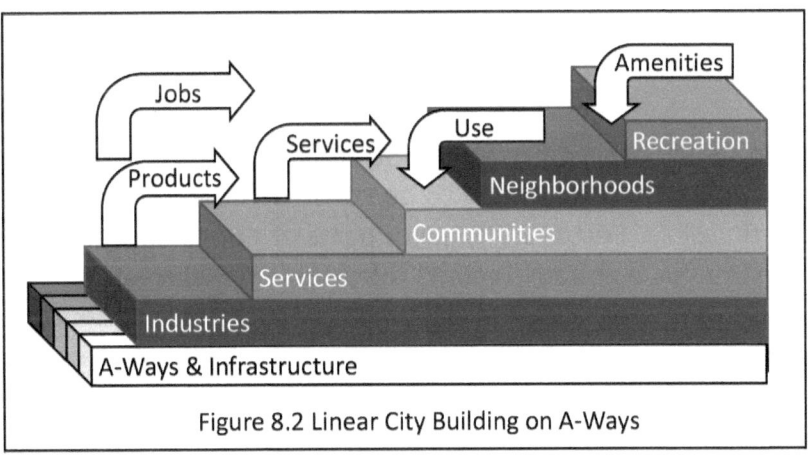

Figure 8.2 Linear City Building on A-Ways

The Emergence of the Linear City

We started the chapter by defining a city as a large concentration of people with an associated complex infrastructure to meet their needs. What we have described thus far meets this definition's criteria since residences, businesses, industries, and community services are all connected together via common transportation and other infrastructure systems. But instead of being configured in a traditional horizontal pattern, we have considered a linear arrangement because our newly designed autonomous transportation system permits us to do so. And this has notable benefits as already described.

Despite this, expansion of these cities must still occur in a horizontal fashion to accommodate increasing populations, services, and industries. Since transportation systems are major drivers of urban expansion, understanding the requirements of a transportation system helps predict the type and direction of growth a city will likely experience. For example, if the transportation system being used involved only bicycles, then growth of a city would be more likely to occur in areas that are flat, smooth, and protected from harsh weather environments. Likewise, based on our previous rule of thumb, growth would be limited to the distance it took to travel about half an hour on a bicycle.

So, what are the requirements of a-ways? A-ways, as noted in prior chapters, have multiple lanes for autonomous vehicles that travel at an array of speeds. Lanes offering the fastest speeds, however, would need to be situated on relatively straight and flat terrains. Ideally, a-ways would therefore need to be constructed in areas where this was permissible. Fortunately, major parts of our current transportation system have already tackled this problem. Railroads and superhighways, and most cities already occupy such terrains throughout the country, so it is likely that a-ways could be built over these existing structures. Why use additional land when existing routes can simply be revitalized with a-way structures?

Once a-way routes are established, vertical development of industries, businesses, neighborhoods, and communities above these structures can then occur all along the a-ways. As cities grow, expansion simply moves along the a-way corridor, and instead of having our current sprawling metropolis in all directions, we now see linear cities evolving in whatever direction the high-speed a-ways are constructed. Instead of having pockets of affluence developing at highway interchanges and railway stations, development would be essentially continuous along a-ways because at any given point, access to transportation would be easy and convenient.

Think about this for a moment. Roads would no longer be required because all transportation would occur via a-ways. This would also eliminate the need for driveways, parking lots and parking garages. All homes, offices, schools, warehouses and other facilities would be located above these a-ways, and therefore, land in between a-ways would not have to be used to support these structures. Once you get rid of all these things that normally comprise urban land areas, you soon see why the design of cities would change. By having a different (and much more efficient) transportation system using a-ways, cities would become more linear in nature and less urban sprawl.

What might your linear city home or office look like? Certainly, the possibilities are numerous and limited only by your creativity. My vision of a linear city involves a dramatic increase in green space. Personally, the view I would like from my linear city home would include a view of mountains, rivers, fields, forests, or even the ocean. I would have an array of trails, bike paths, and lakes where I could enjoy my free time. And when I walked outside my home, I would be minutes from stores, restaurants, a university, and community events because everything would be in close proximity. And my immediate neighbors would be fascinating people with whom I would enjoy working and socializing. The opportunities to connect with others would be markedly enhanced not only by these inherently social environments but by the efficiency of transportation as well.

Of course, not everyone will want to see a mountain or ocean from their home. The great thing about a linear city is that different segments can offer different sights, different specialties, and different activities. In essence, some linear city "neighborhoods" may be more oriented to arts and music while others more aligned with fine dining and cuisine. One may cater to a more active sports lifestyle while another offers greater aesthetics and serenity. The potential to tailor your residence or office to the type of environment you like is not hindered by linear cities. In fact, just the opposite is true. By having more efficient transportation, lower infrastructure costs, and greater proximity to people and services, the possibilities for developing unique environments are increased substantially.

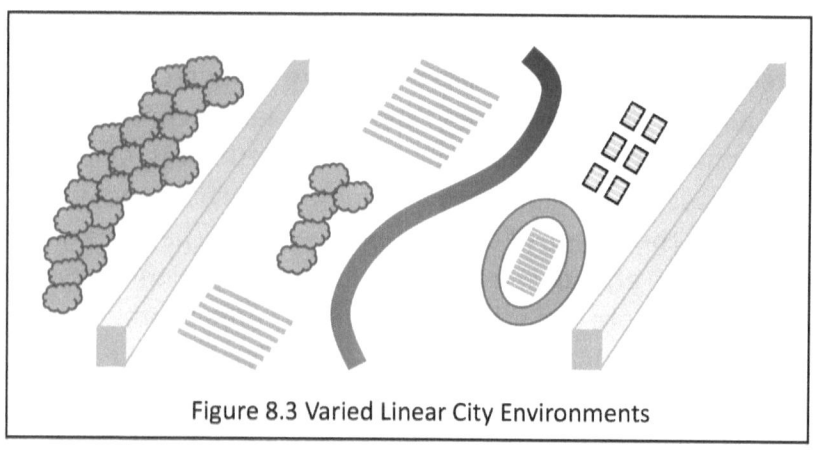

Figure 8.3 Varied Linear City Environments

The Potential of Linear Cities in Real Terms

In explaining the concept of linear cities, we have described how more efficient land use and shared infrastructure will lead to lower costs for everyone. Likewise, with more efficient and accessible transportation using a-way systems, the capacity for serving people's needs will grow substantially. But exactly how much cost savings will actually be gained? And how many more people can be effectively served by such a system? The short answer…a staggering amount. You might be astonished by some actual projections that answer these questions.

Let's start by considering a possible configuration of a linear city. A-ways will carry items and people to their destinations, and situated above these structures are industries, then businesses, then other community services, and then residences. In my dream home, my outwardly facing wall would overlook nature while my inwardly facing wall would allow access to all the services of the linear city community. My floor, ceiling and other two side walls would be adjacent to others (other residences, businesses, etc.).

Given this design, let's examine construction costs of a typical linear city residence. First, we consider the price of the lot. Given that we are sharing space with a-ways and other structures, this cost will be dramatically reduced. Conservatively, this should be half the cost compared to typical land purchases or leases. Secondly, building the basic infrastructure is not necessary since it already exists. Only incremental costs of accessing the existing power, water and waste systems will be needed. These costs should be reduced by as much as 90 percent. Lastly, because of shared walls, absence of driveways and garages, and reduced fees for permitting and inspection, actual construction expenses should drop by at least 80 percent. Put all of this together, and the actual cost of building a home should be less than a quarter of what it costs currently. Those are pretty incredible savings!

The actual configurations of these homes will certainly vary to a degree. Different designs, development constraints, personal preferences, and more will determine what an actual linear city residence might look like. But for the purpose of providing an overview of how linear cities better accommodate people's needs and the use of space, let's make some basic assumptions. For our first assumption, let's presume there are 8 floors of residences in our linear city. Next, let's allot 15 feet of "window" walls facing outward on one side from these homes for each person. With these basic dimensions, a linear city could accommodate 5,000 people per linear mile. By having multiple floors, and by allowing depth of residences between outward and inward walls, these numbers can be easily realized. Additional floors could house businesses and services. For comparison, Warren's father could walk 1 mile inside the Charlestown facility without doubling back, but most of the facility had only 3 to 4 floors. There were 2,700 residents, so the density was 2,700 people per mile.

Let's now consider how we would house these resident populations along our a-ways. At the current time, our national interstate system covers about 50,000 miles of roads throughout the country. If linear cities were built along these interstates, and 5,000 people lived along every linear mile, then this would accommodate over 80 percent of the nation's population. Pretty impressive. But how large are these linear cities? Does just one linear city exist by itself, and if so, how long is it? Can more than one linear city be connected to one another? If so, how does this affect the population being served?

Cities have historically grown to a size limit based on the distance required to travel to the center of a city in 1/2 hour, an average commute of 1/2 hour. Therefore, we can anticipate that linear cities would expand in size based on a similar rule of thumb. In our prior discussion about the configuration of a-ways, and continuous convoys and en-route sequencing, we proposed high-speed lanes traveling at 260 mph, with adjacent lanes traveling at 130 mph, 65 mph, and 30 mph. Based on these speeds, and based on utilizing our continuous convoys at maximum capacities, we can calculate that in 1/2 hour someone could travel at least 30 miles in either direction.

If we assume 5,000 people per mile, then a single linear city could stretch 60 miles and include over 300,000 people, about the population of St. Louis, Missouri. Not only do all of these residents have incredible transportation access allowing them to travel a half-hour or less to their destinations, but the system allows for fast and efficient delivery of all types of materials, supplies, and packages as well. And by housing these within our vertically expanded linear cities, we continue to have an abundance of green space to enjoy!

Based on these calculations, a single linear city could accommodate the population of all but the sixty largest cities in the U.S. If we place a linear city on either side of this linear city a mile apart (in essence 3 parallel linear cities), then people now have access to a much larger area. This essentially triples the number of residents from 300,000 to 1 million. This type of configuration of linear cities would run east and west as well as north and south creating a grid-like structure. Yet, the footprint of the actual linear city structural complex would still be much smaller when compared to our current large metropolitan regions. With the next linear city a mile away, you may not even see it over the trees.

The opportunities for meeting population needs with linear city configurations are quite impressive. Let's take an extreme example. The entire population between Boston and Washington D.C. is roughly 50 million people, and the distance between the two cities is approximately 450 miles. If we designed 25 interconnected linear cities throughout this region of the country, using our assumptions described above, the entire population's needs could be met. But in addition, every single person within this massive area could be in Boston, Washington, or Manhattan in less than 2 hours! This is comparable with the flight time from Boston to Washington not even including travel to and from the airports, baggage handling, and of course security delays. Adding a Hyperloop connecting Boston, New York City, and Washington, DC at 750 mph cuts the long distance travel times roughly in half, so you can get between Boston and Washington, DC in an under hour.

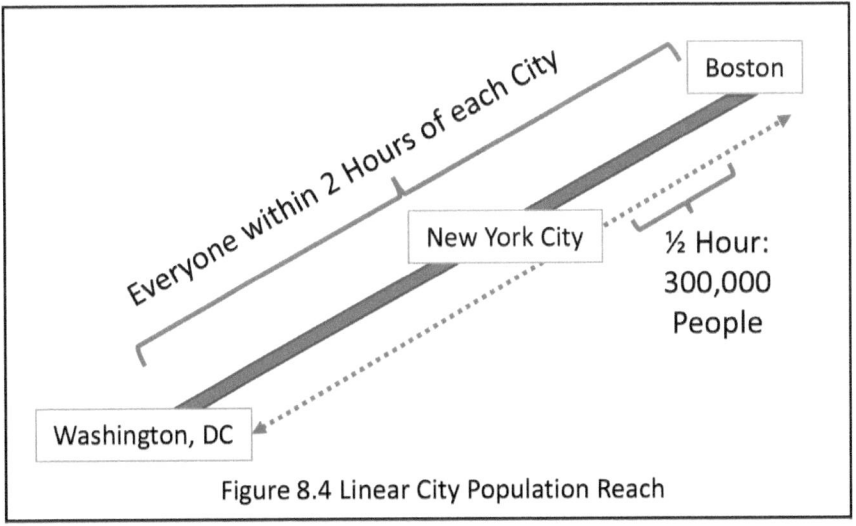

Figure 8.4 Linear City Population Reach

As you can see, the potential changes a-ways allow in urban development are substantial when actual figures are applied to their transportation efficiencies. In addition to encouraging a different design of urban environments (linear cities), they also permit significant increases in the distance a single city covers based on our half-hour commute. Both of these features significantly expand the capacity of cities in meeting population needs. And at the same time, we gain a tremendous amount of green space in the process.

9
C^4 ("See Forth") — Autonomous Navigation

WE CAN STILL visualize the glove box of our cars when we used to travel on family vacations each summer. Within the glove box was everything but gloves...a flashlight, the car's owner's manual, tissues, a screwdriver, and many other very useful but rarely needed items. However, one of the most commonly used things in the glove box was the collection of road maps. In addition to a map of the U.S., every state map from New York to Florida was cataloged neatly in the glove box. When was the last time you actually saw such a map? Pretty amazing how quickly things can change.

Today, we rely on our smartphones for a variety of daily activities, and they are especially useful with our current transportation system. We have a friend who even uses such an application when he travels out of town to "map" his way to restaurants and stores on the same block as his hotel! He can even see the street view of the restaurant to make sure he is in the right place. But as wonderful as these tools are, they have some limitations, and these are readily apparent when we look where our autonomous vehicles will take us. If the world is going to enjoy a truly autonomous transportation system, we all need to take a fresh look at navigation and control.

This chapter provides an overview of how current navigation systems function and the demands required by an autonomous navigation system. In particular, navigating indoors poses additional challenges in comparison to self-driving cars. We describe how a new navigation system will get you from place to place indoors. Navigating in a-ways avoids many hazards faced by self-driving cars, and presents new opportunities to be optimized, including nesting, and continuous convoys. Interestingly, many of the technologies that will go into this new navigation system are already in use, just in different existing systems, and the software is likely to be provided as open source. You can gain a good understanding of how such a navigation system will function by combining these technologies in innovative and economical ways.

Current Navigation Systems

Have you ever wondered how the maps in your smartphone are created, or how they get photos of so many street views? Google and other map providers start with lots of existing map data. But the key to making the maps accurate is paying people to drive along all the roads in the US with specially equipped cars that take detailed photographs everywhere they go. In addition to providing those handy street view photographs, the route the cars follow confirms that you can actually drive there. More people are then paid to look at all this data and create the master map, which provides the data for the map on your smartphone.[72] Tesla does even better than that, providing information in real time. As you drive along, your Tesla is reading the speed limit signs and continuously showing you the current limit on the dashboard, as well as adjusting the cruising speed.

Of course, even after downloading the section of the map you need, your smartphone has to figure out where you are. What you probably think of first for location data is GPS, the Global Positioning

---

72 https://www.theatlantic.com/technology/archive/2012/09/how-google-builds-its-maps-and-what-it-means-for-the-future-of-everything/261913/

System satellites that send signals to your smartphone, so it can figure out your location within a few feet. But with the spread of cellphone towers and Wi-Fi hotspots, there are often faster ways to figure out where you are. Have you ever missed an exit that your smartphone has told you to take? You may notice that the smartphone shows you going along the route it suggested for a while. Eventually you get so far away that your smartphone finally decides you are not following the suggested route and takes a while to calculate a new route. This would be a disaster if that much inaccuracy and delay caused your personal mobility vehicle to crash you into a wall.

As wonderful as smartphone map applications are, they do have some limitations when we begin to consider autonomous transportation. As previously described, autonomous vehicles will not simply travel in a-ways or along roads. They will also travel inside buildings. While some indoor maps are being constructed online for selected stores (like Home Depot), the data needed for indoor maps is not provided by current maps. Likewise, tall buildings in urban environments can interfere with GPS and other location data signals to your smartphone limiting accuracy. And lastly, the resolution provided by GPS systems is measured in feet, which is adequate for navigating current roadways and neighborhoods, but it is hardly accurate enough for autonomous vehicles in tight spaces and within an inch of your destination. These represent some of the challenges related to current navigation systems.

Despite these limitations, self-driving cars are being developed. How do they handle these navigational challenges? For the most part, they use expensive arrays of various types of sensors to assess their immediate environment, in combination with data provided by external location systems. Optical sensors, radar, and laser devices are among the technologies being evaluated. In some ways, these sensors function quite well when traveling on roadways and when parking a vehicle. For example, as you get close to an obstacle, such as a curb or trash can, Tesla shows you how far away you are to the nearest inch. Likewise, sensing obstacles or potential problems is limited to the immediate vicinity and not beyond the roadway. These and other challenges become apparent when we envision all types of autonomous vehicles traveling indoors and around buildings. Therefore, we need to consider how a better navigation model might be constructed to accommodate the specific cost and functionality needs of our proposed autonomous transportation system.

Fortunately, there are other systems is use today that you may not think about as navigation systems. Have you ever played an online multiplayer game such as World of Warcraft? You are navigating in an artificial space, along with perhaps hundreds of other players at their computers around the world. This ability to coordinate locations and motions among vehicles is essential for navigation. Self-driving cars are just now developing the capability to do this type of coordination for nearby cars, but fortunately the technology is already well developed in online games.

Another form of navigation system has been developed over decades for robots. These robots typically use optical sensors, similar to our eyes, to navigate accurately and precisely. And finally, consumer drones are using inexpensive systems to navigate in 3 dimensions. In fact, you can buy drones that include all the elements needed, and the good news for small vehicles is that the processor weighs less than a AAA battery. We will see how these technologies, developed for different applications, can be economically combined to solve the navigation challenges for our autonomous transportation system. The software is likely to be developed open source.

A Typical Journey for an Autonomous Vehicle

Much of the inspiration for developing an autonomous transportation system originated from witnessing the many difficulties Warren's father faced when living in the excellent Charlestown senior facility. So, it seems only fitting to outline some of the major challenges such a system faces in terms of navigation using one of his routine activities…eating in the Charlestown dining areas. Like travelers on a cruise ship, when it is time for dinner in Charlestown, residents descend on the dining rooms for what could be accurately described as a mob. So, imagining how my father might make it to dinner from his room on his autonomous personal mobility device can demonstrate key elements that an autonomous navigation will need to address.

Let's begin with my father in his apartment at Charlestown. He boards his personal mobility vehicle, which is autonomous, and instructs it to take him to dinner. Our first issue is already apparent. How does the autonomous vehicle know where "dinner" is? As we have already noted, typically neither car nor smartphone map applications have information on the inside of buildings or campuses. Thus, our first issue is how to provide our autonomous vehicle with this data so that it can navigate.

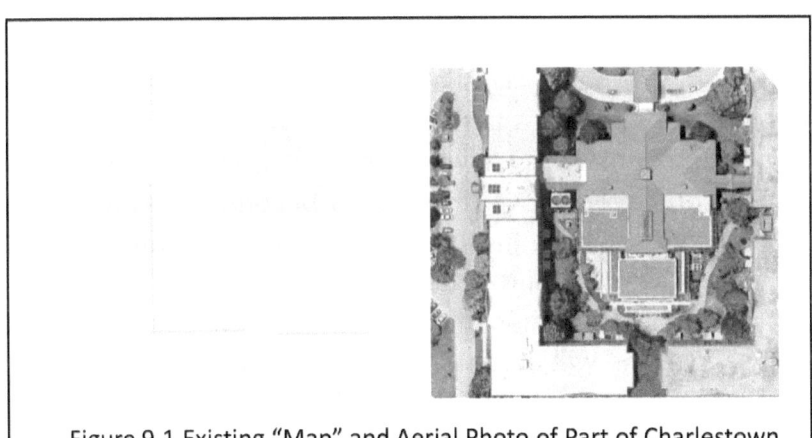

Figure 9.1 Existing "Map" and Aerial Photo of Part of Charlestown

Assuming, for the moment, we address the problem of providing map data about the location of the dining area in relation to my father's room, next the autonomous vehicle must determine the best route to take and navigate its way to the dining area. How far should it travel down a hallway before turning? Which direction does it need to turn once it gets there? Which route is most efficient? How will it know if a hallway has been closed for construction? We may be used to addressing some of these issues on the road, but we have never had to address them within and between buildings. And what if the dining room is on a different floor, where are the elevators? Maps would help, but even with a map, there are other navigation issues.

At this point, we will assume my father's autonomous vehicle is traveling down hallways and turning corners guided by a new navigation system. However, along the way, it begins to encounter other residents walking up and down the hallways, and some have leashed pets that scurry along with them. Likewise, my father's vehicle encounters other autonomous vehicles carrying residents as well as supplies and deliveries. What will prevent his autonomous vehicle from crashing into someone or into another vehicle? This is no longer just a transportation issue…it is a safety concern for all involved. Not only will our navigation system need to recognize fixed objects, but it will also need to recognize moving objects as well.

Finally, my father arrives at the dining area. While traveling from my father's room to the dining area was challenging in terms of negotiating doorways, hallways, and elevators, the need to move even more precisely in the dining area becomes readily apparent. The autonomous vehicle must somehow move between tables and chairs, dodge waiters and other patrons, pause and start at the appropriate times, and then place my father within an inch or two of the table where he will dine. All of these activities require considerable accuracy to successfully get my father to his destination. As you can appreciate, this is well beyond the demands of current GPS navigation systems and self-driving cars.

At first glance, you might assume the ability to address these issues is going to be expensive and far in the future. But in making this assumption, you would be mistaken. Developing a new navigation system for autonomous transportation is perfectly feasible using many of the technologies we take for granted today. In the next few sections, different components of such a system will be described so that you can understand how an autonomous vehicle can navigate all of the challenges and obstacles just described. In doing so, you will be able to see that autonomous navigation is not only possible but a real option for modern transportation systems.

Autonomous Vehicles and Navigation Systems

One approach to building a map of the inside of Charlestown is for my father's personal mobility vehicle to just start looking around his apartment. After all your Roomba vacuum cleaner just zooms around and covers all the floor in your home. The optical sensors in my father's personal mobility vehicle can do a much better job of figuring out the walls, doors, and furniture. In fact, that's probably the best way to map his apartment. How else would it know where all the furniture is, especially since chairs and other pieces can be moved around.

But how would his vehicle know about walls, doors, furniture, and other common items. After all it took each of us years as children to learn how to name and navigate around these objects. Fortunately, computers can just have all the needed information loaded at once. Each vehicle will come equipped with a basic vocabulary and model of what all these items are, and what they look like. That way when my father says take me to my easy chair, the personal mobility vehicle knows what an easy chair is, although the first time my father would need to indicate which is his easy chair, to distinguish it from my stepmother's. Knowing how to go through a door, and how to move close enough to a table so my father can comfortably eat, will save a lot of confusion and even some bruises.

So, what about getting to the dining room? My father could just give his personal mobility vehicle step by step instructions: go out the door of his apartment and turn left, then right at the next intersection, and the third doorway on the left is the dining room. He can name other things along the way: the door to the library, his friend John's apartment, and the elevator. His personal mobility vehicle will be using its "eyes" accurately recording the details along the way and turning all that information into a detailed map. This would indeed produce a detailed map, but slowly, and only for the areas that my father went through.

There is a better way. When my father moved into Charlestown, he received a paper map of the entire complex. Extending this model, Charlestown can also provide a digital version of the map. This allows my father's personal mobility vehicle to immediately know the location of all four dining rooms, their names, and how to get there, along with all the other locations and routes. While the paper map of Charlestown is 2-dimensional, the same as your smartphone maps, the digital map of Charlestown needs to have more detail. Not only do the dimensions need to be more precise, for example will my father's personal mobility vehicle fit though the door to his clothes closet? But we need to include the 3rd dimension, height. For example, will may father's legs fit under the dining table? You might be nervous about having a vehicle autonomously getting you that close to a table, and you might prefer to do it yourself. However, recall I noted that my step-mother didn't have the necessary fine motor skills, so she, at least, needs the motion performed autonomously.

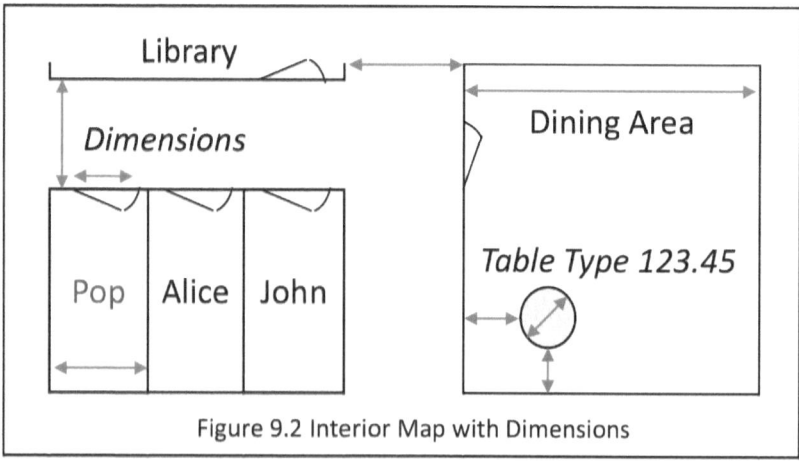

Figure 9.2 Interior Map with Dimensions

In addition to recognizing these objects the pre-loaded information will include models of how they move. For example, an adult walking along will go about 3 miles per hour mostly in the same direction, but faster when running, while a dog is likely to dash off in a different direction without any notice. And of course, there will be instructions not to crash into people or pets or other objects. There are always tricky cases, such as a child darting in front of a moving vehicle — nobody said this would be easy. That's a key reason we invented a-ways, to separate pedestrians and pets from fast moving vehicles.

All of this may sound complex and expensive, and you wonder if it is feasible. Consider your smartphone — actually that's a misnomer! It's really a computer that fits in your pocket and happens to have a phone app. But it also has a camera app, a video app, a map app, and many more. You can talk to your pocket device, and it understands and does what you ask ... well sometimes, but it is getting better. This computer that fits in your pocket has ample processing capacity and memory to do the tasks we have just discussed.

In addition, your smartphone has a camera which can capture what is going on, and process that information, just like we described as needed by an autonomous vehicle. For example, have you noticed when you are taking a photograph with people in it, the screen has outlined the faces of the people? Some new "smartphones" even have stereo cameras so they can detect distances, just like the robot vision systems. So, the device that fits in your pocket already has the capability to do all the sorts of things we've just described, and much more, so it won't be a problem for autonomous vehicles. Even better the device integrated into autonomous vehicles will be a lot cheaper because it does not need the display screen, the case, the separate battery, and other expensive components.

Cloudlets and Autonomous Navigation

My father's personal mobility vehicle is whisking him to the dining room, avoiding obstacles in the hallway, and everything is working safely. But what happens if his autonomous vehicle is about to travel around a corner, and another autonomous vehicle is approaching?

Because the corner obscures sensor detection of the other autonomous vehicle, a crash may occur. How can we see around a corner? What we need is information about other nearby vehicles, even if they are out our line of sight.

One approach is to have each vehicle send their information into the cloud, where others can access it. But we need this information right now, seconds count. What if the network access is busy or down? Safety is too important to risk those delays. Fortunately, this information is only needed for things nearby. If you are across town you don't need to know about someone moving down the hall in Charlestown, in fact this is private information and you shouldn't be able to see it — we'll talk more about privacy in this and later chapters.

Because this information is only needed locally, it makes sense to send it directly between nearby autonomous vehicles. But how do we organize this information sharing? Rather than sending the information into an all-encompassing cloud, we will form a local "cloudlet" of just the nearby vehicles that actually need the information. Nearby autonomous vehicles will communicate securely with each other using Wi-Fi or other local communications alternatives to share their perspectives to add information to a local map. This shared information will include everything relevant to navigation in the area, including furniture, people and pets.

This sharing is even more important as my father approaches the dining area. There are people leaving and entering the dining area, people talking and standing in line inside, and people moving to and from their tables. Perhaps another dozen autonomous vehicles are in the dining area, and all are joined together in a cloudlet sharing their sensor data with each other. With all the people and other obstacles any single autonomous vehicle cannot see the whole situation. But the additional data shared with the other autonomous vehicles in the cloudlet allows them to create a comprehensive map because of all the additional perspectives available. Cloudlets thus facilitate data sharing and the creation of more accurate and detailed maps for navigation. So before getting to the dining room, my father's personal mobility vehicle joins the dining area cloudlet and gets a current map of the area, with all the action.

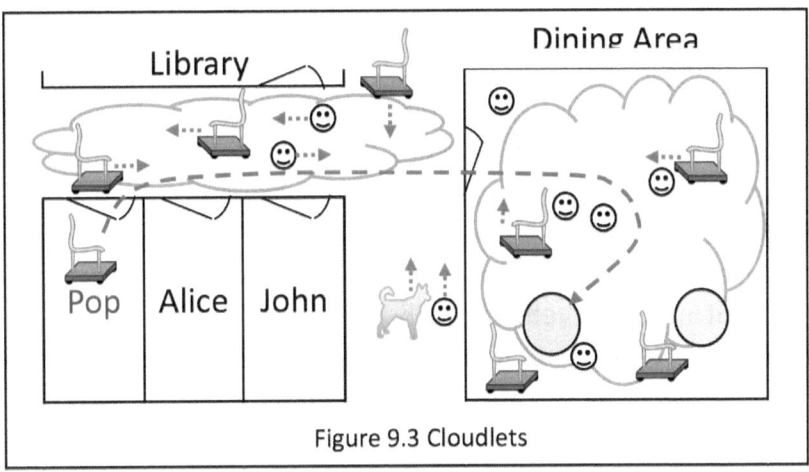

Figure 9.3 Cloudlets

You may be wondering whether the amount of information to be shared is too large for the communications capacity or for the autonomous vehicles to process. For example, do they need to share videos of everything that is happening to predict where people and other moving objects will be? Recall that we said the autonomous vehicles would already have models of the objects they are likely to encounter. These models will also include the information needed to predict their motion. A few numbers are sufficient to allow autonomous vehicles to predict how each object is moving. Again, toys have already solved some of these challenges, for example, the Microsoft Kinect game console already senses people and models their motions in detail.

This may sound complicated, but using our two eyes, our brains are constantly making these predictions without even consciously thinking about it. When the batter hits a high fly ball, the outfielder is already running to where the ball will come down while the ball is still going up. And you do this every time you cross the street, deciding whether you can safely get across the street before the oncoming car gets here, or you should wait, or perhaps you need to hurry to make it across safely. Autonomous vehicles can do these same kinds of calculations using a pair of optical sensors, like our eyes, and just need a few numbers to accurately predict the motions. Another advantage of sharing information is that smaller vehicles can get help from larger, more capable vehicles to solve challenging navigation situations. For example, a convoy vehicle can provide navigation assistance for small vehicles inside.

What happens to the map of a local area when there aren't any autonomous vehicles in an area? Is all that carefully assembled information lost? The static information is already in the map that my father's autonomous vehicle received from Charlestown. Part of the navigation system is an infrastructure to supply these static maps to authorized users. My father's autonomous vehicles can retain the static information he has collected from frequently visited areas, including the names he has assigned to objects and areas.

But what about the dynamic information? People and pets moving around are definitely dynamic. The dining area may be completely rearranged, for example at Halloween, including moving all the tables and adding displays that change the local map. So how can we store this information, and even include changes? Consider what other useful devices may be in the area: personal computers, security cameras, and smartphones all have cameras and processing capacity, so they could both store the local map, and even keep it current. These devices can also supplement the processing and sensing capability of smaller vehicles, allowing us to make all the components simpler and cheaper, while improving accuracy and safety.

C^4 ("See-Forth") — Cloudlet Communication, Computing and Control

So far we have focused on the local aspects of navigation, using communications and computing in a local cloudlet. But navigation also needs to address the bigger transportation picture, that is traveling longer distances, and at higher speeds. Fortunately, these same techniques apply to this larger situation, and we just need to add another dimension ... control. The term control in this case refers to the ability to manage traffic and efficient transportation of the entire system. We will describe how information created in the local cloudlets can be aggregated to drive the control processes. Thus, we have named this navigation system C^4, appropriately pronounced "See-Forth", both because there are four C's, Cloudlet Communication, Computing, and Control, and because this is a 4-dimensional system, where the fourth dimension is time.

Applying this to the Charlestown dining areas, think about the data needed to manage these operations. What time should each dining area open and close? How many servers should be on duty at any time to provide good service while also being economical? The dining area cloudlets can collect the statistics needed to answer these questions and many more. We should emphasize that these are statistics, not the travel data for any individuals — maintaining privacy is a key aspect of providing this data. Aggregation is an essential technique to achieve this important goal. These are historical prediction aspects of control. There are also real-time aspects, for example routing people around a corridor where the carpet is being replaced, or letting people know that the Tower Dining Area has a 15-minute waiting line now, while the other three dining areas have no lines.

For more conventional traffic control, you are probably thinking of stop lights, signs, and police directing traffic around crashes and construction. The traffic displays on your smartphone maps are another aspect of traffic control, sometimes the maps will even direct you around traffic delays. With autonomous vehicles we have much more effective methods of control to optimize traffic flows and speed you on your way. Statistics and other aggregated data are collected by the local cloudlets and passed up through successive layers of a hierarchy, effectively larger cloudlets, to manage traffic in larger geographical areas. Nesting of smaller vehicles inside larger ones provides extensive opportunities for optimizing both routing of individual vehicles, as well as scheduling larger vehicles and convoys to virtually eliminate delays.

Today when you are planning your route to get to work or to take a trip, your plans for that trip stay with you. What if the statistics of that trip, including your planned departure time, were aggregated and used to optimize your whole route? The C^4 systems can optimize your actual route based on everyone's actual plans, rather than just historical averages. The C^4 systems can also optimize the schedules of the convoys to eliminate delays for you ... all while maintaining your privacy. In a later chapter we will consider many more aspects of privacy, and the opportunities for secure services, such as deliveries, enabled by autonomous transportation

One of the key reasons these improvements are possible is routing high-speed travel through a-ways. This assures protection from all sorts of hazards: weather, animals, debris, pedestrians, and human drivers, eliminating crashes and other delays. A-way systems will have highly integrated communications and data sharing capabilities to both collect data and precisely manage the timing and flow of a-way traffic. Another reason is that all the electronics, information, and communication technologies will be located within the a-way structure itself. These systems are protected from outside elements while also being readily accessible for maintenance and repair. And these infrastructures allow constant monitoring and surveillance of a-ways for enhanced safety and functionality of the overall transportation system.

In this chapter we have constructed our autonomous navigation system from the ground up (or at least from our autonomous vehicles up!). We have addressed the issues to get my father from his apartment to a dining area at Charlestown safely and quickly. The novel technique of local cloudlets enables sharing of data among nearby autonomous vehicles to overcome the problems of limited line of sight. Fortunately, existing technologies and techniques form the basis for this new navigation system, so we anticipate rapid development of economical solutions, even for the smallest autonomous vehicles. Because the basic navigation is local to the cloudlets, this software is likely to be produced open source. The navigation system also enables optimization of travel, both locally and everywhere through high-speed a-ways for fast, safe and efficient transportation.

10
Sustainability

WARREN: FOR MANY *years when I was young, my family and I would visit my grandparents in Florida. Because of school calendars and other commitments, our trips would always occur in the summer months. For those of you who live in Florida, you might think this is not the best time of year to visit the state, especially since the wiser "snowbirds" purposefully schedule their visits to Florida between Thanksgiving and Easter. But in actuality, our visits were always quite comfortable from a climate perspective despite my grandparents not having air conditioning. In addition to having natural trees and landscape to shade their home from the sun, their house had porches all around, and vaulted ceilings with a cupola at the top to allow hot air to escape. I don't ever recall being hot or uncomfortable in their home during our visits. These reflect some of the natural strategies for sustainability through design that in some ways we have lost.*

While different definitions of sustainability may exist, one perspective of sustainability can be described as making the best use of what we have without wasting our inherent, natural resources. Comparing many homes to the old homes, you may appreciate that many subdivisions were constructed after the land was cleared of its natural environment to make room for the new homes. Walls and roofs were put into place to create protections against the sun, but in the process, available energy from the sun was lost. And small shrubs and saplings were planted to replace the trees removed offering little if any shade. Then we burn expensive fossil fuels to heat and cool these homes. Based on our definition above, these practices are hardly sustainable. We haven't necessarily made the best of what we had, and we have been wasteful in the process.

Many opportunities exist today for us to dramatically improve sustainability of our resources, and this is certainly true when it comes to transportation systems. Such opportunities are apparent when considering new designs for vehicles and infrastructures, but likewise, other areas for enhanced sustainability also exist through a better transportation model. From water and waste management to environmental preservation, the autonomous transportation we propose offers significant advantages in resource sustainability and enhanced efficiency. While some of these topics will be considered in greater detail later in the book, this chapter provides an overview of these advantages and opportunities. As you gain an understanding of these potential benefits, you can appreciate why a new transportation system is not only a good idea but one of real necessity.

Enhanced Energy Efficiency

You would probably not be surprised to know our current transportation system needs significant improvements when it comes to energy efficiency and sustainability. But did you know that transportation consumes 28 percent of all the energy we use as a nation? Therefore, any improvement that we could make in transportation energy efficiency could have a major impact on energy usage and sustainability overall. Is this possible? Absolutely! Our current transportation system is the least energy efficient sector at 21 percent.[73] In other words, nearly 80 percent of the energy used for travel is wasted.

---

73 U.S. Energy Information Administration. "Total energy." Website, 2017. Retrieved from https://www.eia.gov/totalenergy/data/annual/showtext.php?t=ptb0105

The energy efficiency of current transportation systems could be improved by at least 75 percent, and our proposal for autonomous electric vehicles and a-ways goes a long way to achieving this goal. By reducing the weight of vehicles from two tons to less than the weight of the load they carry, the energy needed to move us (and other items) around falls substantially. By using electrical energy to power our vehicles instead of gasoline, we improve energy efficiency from 25 percent to over 90 percent, and we can obtain the electricity from the sun and wind. These two features alone create a situation where energy sustainability is greatly enhanced. And in the process, we become less dependent on oil while promoting a healthier environment.

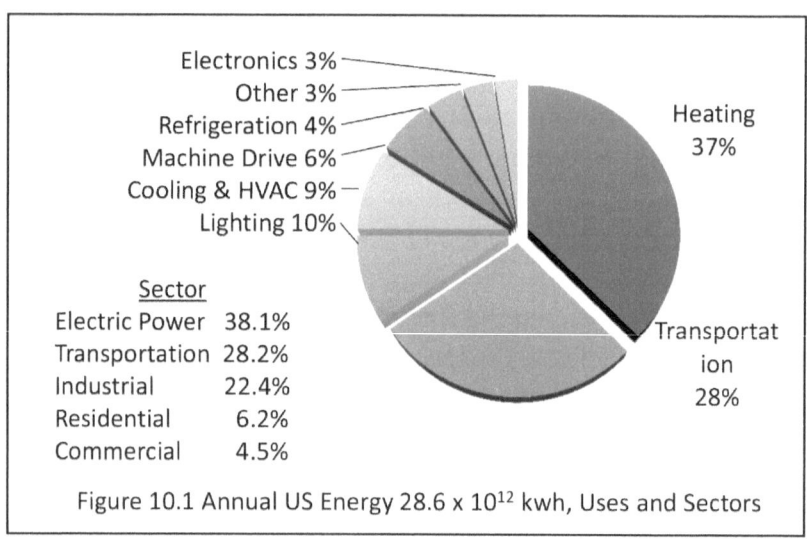

Figure 10.1 Annual US Energy 28.6 x 10^{12} kwh, Uses and Sectors

By having autonomous vehicles that are lighter in weight and powered by electricity, energy requirements are naturally less, but what about recapturing some of the energy used by these vehicles. One of the areas where gasoline-powered cars are highly inefficient relates to low-speed transportation settings. In general, the lower the speed, the less energy efficient gasoline vehicles are. This is why "city MPGs" are always less than "highway MPGs" for our cars. When gasoline-powered vehicles come to a stop, the existing kinetic energy of the car's motion is suddenly lost to the surrounding environment as heat. In contrast, autonomous vehicles (as well as autonomous elevators) use regenerative braking systems that allow the kinetic energy of movement to be recaptured and used to help power vehicles.

Autonomous vehicles are not the only source of enhanced energy efficiency in our new autonomous transportation model. A-ways also promote greater sustainability of energy through their inherent design. Consider the road surface within a-ways. Unlike current roads that are exposed to the elements and constantly in need of repairs, a-ways are protected due to their enclosure. This produces a much smoother road surface with less friction, and less friction means less energy required to propel autonomous vehicles along their paths. This aspect of a-ways further reduces overall transportation energy requirements.

Of course, being inside an a-way has other advantages when it comes to sustaining energy resources. These enclosures protect our vehicles from being exposed to heat and cold, rain and snow. Unlike today's cars which require protection from the harsh environment and need built-in heavy-duty heating and air conditioning to create a comfortable driving environment, autonomous vehicles would enjoy a much milder environment. Thus, autonomous vehicles in a-ways require significantly less energy.

With this in mind, let's talk about sunlight for a moment. Did you know that the sun provides us with about 1,000 watts of energy for a square meter? Allow us to put this into perspective. You would need 109 conventional 60-watt incandescent bulbs to create this much useful light, and you would use 6,540 watts of electrical energy. Even if you used LED bulbs, you would still use 1,627 watts of electrical energy. So just 1 square meter of sunlight could light your whole house if you just use it directly. Surprisingly, sunlight is more efficient too, so it adds less heat to your house than all those bulbs.

If we could completely capture the total energy that the sun showers on the US, it would only take less than one minute of sunshine on a summer day to meet the energy demands of the United States for an entire year. Unfortunately, the best current solar systems only capture about 20 percent of this power with additional losses of efficiency occurring when electricity is transmitted, stored, and used. Regardless, the potential for enhanced solar systems exist, and a-ways could use these new technologies directly. By constructing a-ways and linear cities using such materials, solar energy could provide the total needs for heating, for lighting, as well as for electricity to support autonomous vehicle transportation.

Warren: You might think this sounds too futuristic to be a realistic consideration, but that would be a poor assumption. Recently, I visited an exhibit by a remarkable artist named Stephen Knapp who creates vibrant, stunning displays of colored light with the use of light filters and prisms.[74] I was so impressed with his work that I began exploring how the separation of sunlight spectra might be used to provide us with light and other energy needs. In the process, I learned that a little more than half of the energy of natural light is infrared in nature, and this portion could be used directly as heat for a variety of applications.

About one-fifth of the sunlight, however, could be used directly as visible light. Instead of having to generate electricity inefficiently and then turn it back into light inefficiently, as we saw from the lights in the average home today, sun light could be distributed directly much more efficiently to put light when and where we need it, using autonomous systems. At night, we could use very efficient light sources as input to the same light distribution system.

While these technologies are not yet developed for use in our homes and in a-ways, their potential for use in the near future is tremendous. And unlike current road systems that would pose many challenges in using such technologies, a-ways provide ideal structures in realizing their benefits. From autonomous vehicle construction to a-way designs,

---

74 Stephen Knapp. "Stephen Knapp." Website, 2017. Retrieved from http://www.stephenknapp.com/

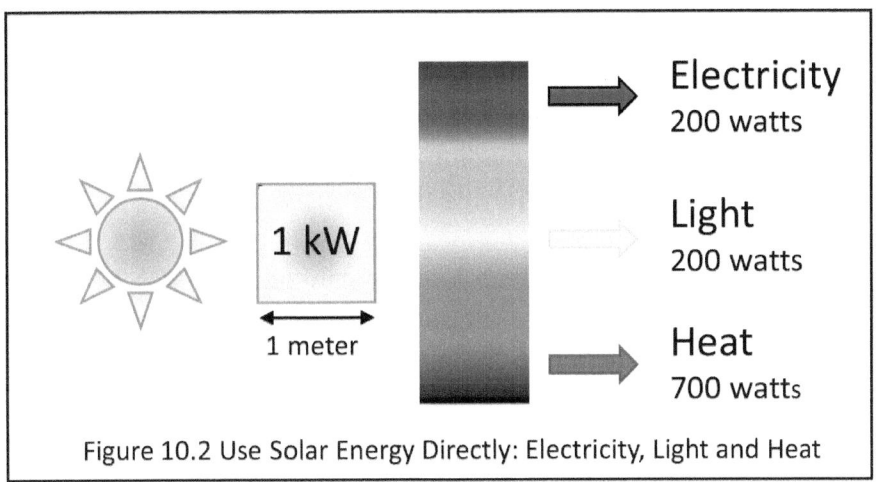

Figure 10.2 Use Solar Energy Directly: Electricity, Light and Heat

opportunities to greatly enhance energy efficiency and sustainability are obvious. This pertains not only to today but to the future as well.

Waste Management and Sustainability

It is a beautiful sunny day, and you have decided to go to the beach. You are looking forward to the fresh, salty breeze and soft sand between your toes. As you leave your house and head down the expressway with the windows down, you suddenly catch a whiff of some wretched smell. Did you leave a bag of trash in your car? Perhaps you forgot to take the leftovers from the restaurant where you dined last night into the house. But then you look over to your right and see what looks to be a large hill several hundred yards away with dozens of birds circling the hill overhead. It is then you realize the odor is not coming from your car but from the landfill that seems to be growing in size (and smell) with each passing year.

Most of us are aware of the issues surrounding sustainability and waste management. This awareness is a good thing in that the number of Americans who recycle has steadily grown. But you may not be aware of a few key statistics. Did you know that the amount of municipal waste generated each year in the U.S. now exceeds 250 million tons?[75] That is a lot of trash! Of this amount, a little more than a fifth can be attributed to paper and cardboard, and only 40 percent of these materials are being recycled.[76] In addition to reducing the volume of waste, and providing material for other uses, recycling saves greenhouse gases equivalent to those emitted by 39 million cars![77]

Waste management is thus a major concern for each of us. Not only does it make a negative impact on the aesthetics of our landscapes, but it also damages the environment while depleting available resources for the future. However, a tremendous amount of this waste is completely unnecessary. Have you recently received a package in the mail only to find that the amount of packaging and related materials takes up ten times as much space as your actual item? Where does all that packaging go after you are finished? Probably to the same landfill or recycling center we just mentioned.

So how can our new autonomous transportation system help with waste management? In many ways, actually. We will discuss the potential for this system to help with water management in a later section, but suffice it to say the system can help eliminate all those non-reusable plastic water bottles that currently litter our streets, playgrounds, rivers and oceans. In addition, our system can also eliminate the use of non-reusable packaging materials. In doing so, the percentage of current waste attributed to paper and cardboard could shrink dramatically.

In our prior discussions about autonomous vehicles, we described how medications and other items could be delivered "just-in-time." Each delivery autonomous vehicle would have its own mobility platform and its own reusable container, which would be designed according to the

---

[75] U.S. Environmental Protection Agency. "Advancing Sustainable Materials Management: 2013 Fact Sheet." Website, 2015. Retrieved from https://www.epa.gov/sites/production/files/2015-09/documents/2013_advncng_smm_fs.pdf
[76] Ibid.
[77] Tierney, Jon. "The reign of recycling." The New York Time, 2015. Retrieved from https://www.nytimes.com/2015/10/04/opinion/sunday/the-reign-of-recycling.html?mcubz=3

contents that it carried. Let's take this a step further. In addition to the overall dimension of the container being the correct (and smallest) size, the container itself would protect the contents from being damaged. This might require physical cushioning in some cases, temperature control in others, or some other type of design unique to the item's needs. In any case, the materials used would not be discarded after the item was delivered, but the entire container would be moved for cleaning by an autonomous vehicle, and then moved for re-use.

By having autonomous vehicle reusable containers transport materials, there would no longer be a need for packing peanuts, cardboard boxes, packing tape, and many other wasteful materials. For those of you who like to "pop" the bubblewrap, this might be a little disappointing. But in relation to sustainability, the impact autonomous vehicles could make in this regard could be tremendous. Instead of such materials contributing to our municipal waste totals, autonomous vehicle containers would permit repeated use without waste. The only additional requirement in using these containers would relate to appropriate cleaning between deliveries. As we will discuss later, these containers could even be used to help with recycling, trash and other waste elimination for which we currently use trucks, pipes and other infrastructures.

Options for the exact reusable materials included in autonomous vehicle containers will likely evolve as these systems are put into place. Malleable, durable, and cleanable materials may be developed to meet content requirements in some cases while insulated designs might be used in others. Regardless of the actual materials used, these would no longer contribute to the volumes of daily waste seen today with packaging materials. Of course, autonomous vehicles would inherently create less waste themselves when compared to current vehicles (no oil or antifreeze leaks), and enclosed a-ways would provide cleaner environments with less dirt and debris. But the impact reusable containers in this autonomous transportation system would make on waste management may likely be the most impressive of all.

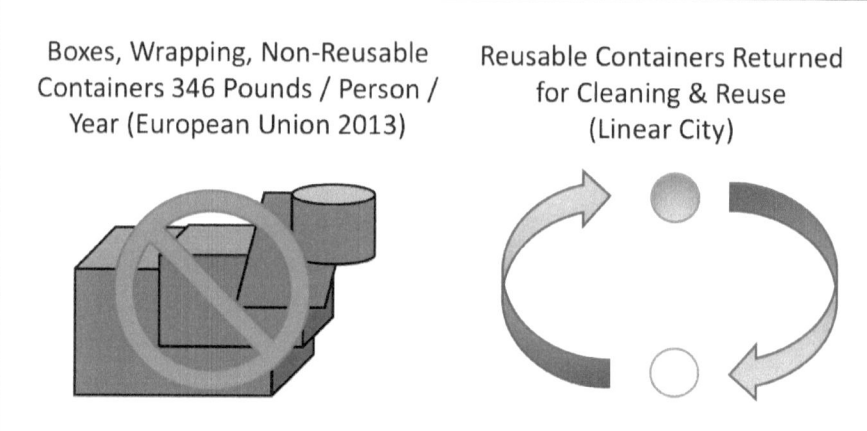

Figure 10.3 Sustainable Reusable Containers & Efficient Transportation

Infrastructure and Sustainability

Warren: One morning not long ago, I was kayaking and got an alarm that our house power was out. Later our neighbor told us a large truck was mistakenly driving down our street. The poor driver was lost, since negotiating my neighborhood in such a vehicle would not have been something he would have wanted to do. He turned into our neighbor's driveway and his truck destroyed the power lines and the fiber optic lines to our whole neighborhood. There was nothing the driver could do at that point except sit and wait...literally. You see, his truck was draped in power lines at that point, and he could not safely exit his truck until the power company arrived to disconnect the power. And of course, the neighborhood was left without power, Internet or phone service.

Your first thought might be that the simple solution would be to remove the above-ground wires and poles so such situations never occurred. But these structures were placed above ground for logical reasons. For one, access to these structures is important when emergency power outages occur, when phone lines need to be repaired, or when television cables need attention. But at the same time, this location places these structures at risk for weather-related damage as well as for problems that might occur with lost truck drivers. This is hardly a system that promotes sustainability.

In many communities, however, infrastructures have been placed underground. This notably eliminates the hazards just described from vehicles and weather, but it does not prevent human-related encounters altogether. Every year, various pipes, wires, and cables are damaged from homeowners and construction workers digging in the wrong place. Likewise, utility crews spend significant amounts of time spray-painting sidewalks, curbs, and easements to identify where underground structures are located. And when repairs are needed, work crews have little choice but to dig up existing roads, walkways, and lawns to address the issue. Here again, resources are not necessarily being used in the best way possible.

You may not realize that our transportation infrastructure similarly affects our ability to effectively manage waste. Construction and demolition waste each year exceeds 530 million tons.[78] That is twice as much as our municipal waste! Of this construction waste, more than 40 percent is related to work and repairs on roads and bridges.[79] Reducing these repairs in both frequency and magnitude will be important in pursuing greater sustainability. However, many infrastructure areas are facing serious problems in this regard as our nation's infrastructure ages and technologies become outdated.

It's no secret that our nation's current infrastructure has critical problems. The American Society of Civil Engineers has rated our nation's overall infrastructure a dismal "D+" overall. They estimate $2 trillion in additional investment is needed over the next 10 years, increasing a full 1% of GDP to 3.5%. Many areas are facing serious challenges as outdated pipes and structures require massive investments for repair.[80] Even worse, road and other infrastructure

---

78 U.S. Environmental Protection Agency, 2015.
79 Ibid.
80 American Society of Civil Engineers. "Infrastructure report card 2017." Website, 2017. Retrieved from https://www.infrastructurereportcard.org/

repairs cause transportation delays, safety hazards, and the inefficient use of human labor. Plus, when these repairs are finished, rarely is the road surface smooth and pleasant to travel. There has to be a better way.

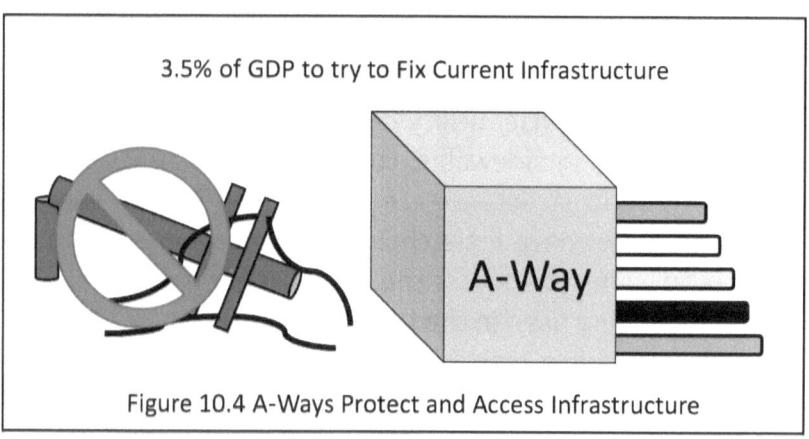

Figure 10.4 A-Ways Protect and Access Infrastructure

With the use of a-ways, most of our sustainability problems can be addressed more effectively. Within these enclosed structures, electrical lines, water pipes, sewage lines, and telecommunications cables could connect services to communities. This immediately resolves some of our infrastructure problems. By having these structures inside an enclosed a-way, weather, water, and animals will not be able to cause problems or malfunctions. Likewise, these infrastructures will be in a location where accidental damage by trucks, homeowners and construction crews can't occur. So far, we have already significantly improved our situation.

In addition to these improvements, an enclosed a-way also allows easy access to these infrastructures for installation, repairs and maintenance. Maintenance is inherently less costly than repairs, and being able to do this without digging up pipes and wires requires less effort and use of resources. Likewise, repairs can be performed faster due to ready access and the potential to quickly replace "segments" of an infrastructure with functional ones while the removed segments are being repaired elsewhere. Lastly, this type of a-way design allows newer technologies to be adopted more easily than our current systems. The American Society of Civil Engineers has determined the costs involved in repairing and replacing our infrastructure with existing technologies are beyond our abilities from a financial standpoint. Thus, the cost savings and other benefits highlight the need for new technologies provided with a-ways.

All of these sustainability advantages cited with a-ways are further amplified through linear cities. In linear cities, the infrastructures located within a-ways can be easily accessed by industries, offices, homes and communities that are located immediately above these a-ways. The distance to access the infrastructures is greatly reduced when compared to current infrastructure systems, and this will further reduce the number of problems that might occur overall. This type of design can also facilitate resource sharing. Since linear cities allow close proximity of residential, commercial, industrial and municipal areas, better resource distribution can occur at lower costs.

At this time, we simply cannot afford to fix the nation's current infrastructures by replacing them with the same types of structures with old, outdated technology, and it makes little sense especially from a sustainability point of view. Rather than digging up existing structures, and trying to repair our roads and bridges, it is only feasible to consider new options that allow newer and more efficient technologies to be used. A-ways and linear cities provide such an option through designs that embrace sustainability, along with the many other benefits.

Autonomous Transportation and the Environment

For many of us who live in major urban areas, the way we use transportation might seem pretty odd to the outside observer. During the week, we sit in traffic every morning as we commute from the suburbs into work downtown. Then, in the evening, we do the same thing in reverse. When it comes time for the weekend, and when we take our vacations, we often drive long distances (and waste considerable amounts of time) to escape the over-built and over-paved areas we have created and seek out various natural environments for our recreation. And even those who have decided to live downtown to avoid the daily commutes, must travel outside of the city to reach green space destinations.

When it comes to the environment, sustainability, and transportation, major issues exist concerning air pollution, water pollution, noise pollution and climate change. Any transportation system that reduces greenhouse gas emissions can certainly be seen as a step in the right direction, regardless of your political views and beliefs concerning these issues. So, rather than making the argument about how autonomous vehicles and a-ways are much better for environmental sustainability based on these parameters, we are going to take a different approach. Let's simply consider the enhanced sustainability of green space (and space in general) with the autonomous transportation model we have described.

As a starting point, first consider the amount of space our current roadways, driveways and parking lots consume. As previously noted, if we added together all the paved areas in the country, we would cover a land area larger than the state of Georgia to accommodate its size. If we consider the national interstate system alone, it covers over 48,000 miles itself. Therefore, any transportation system that offers a smaller footprint, yet enhances transportation efficiency and functionality would be quite attractive, right? This would be even better if it accomplished these tasks while allowing us to dramatically reduce our costs of transportation. That is exactly what our autonomous transportation system does.

Beginning with autonomous vehicles, we can already start to appreciate how much space we can save. If you recall, our autonomous vehicles not only weigh much less than today's average vehicle, but they also require much less space in general. By the time various features such as bumpers, engines, and other components are eliminated, autonomous vehicles that carry human passengers are substantially smaller. If we are considering other items and packages, autonomous vehicles are uniquely designed according to their contents' needs. Here again, wasted space is minimal. And through nesting techniques where various containers share the same autonomous vehicle, space is again maximally utilized.

A-ways offer additional advantages in terms of space preservation and use. With smaller autonomous vehicles and nesting strategies, a-ways naturally occupy much less space than a traditional road. Continuous convoys also help with this since autonomous vehicles, in essence, are nested together within these convoys, and convoys are linked together and well-controlled through automation. These features mean fewer lanes are needed within a-ways in addition to fewer "roads" in total. By making some gross calculations based on average size and speeds of these autonomous vehicles and Continuous Convoys, the space required by a-ways for the same transportation demands would be 15 times less than our current road systems! That is pretty remarkable, especially when you consider a-ways would get you to your destination faster and autonomously.

Now, we come to linear cities. While we enjoy the space-saving capacities of tall buildings and skyscrapers today, the opportunities for us to build vertically on top of a-ways far exceeds what we have realized thus far. Because all infrastructures can be housed within a-ways, the ability to develop cities situated above these structures is greatly enhanced. In doing so, we get the best of both worlds. We have important amenities at our fingertips (including autonomous transportation and robust infrastructure), and we have immediate access to green spaces for our recreation and vacations. In addition, autonomous elevators greatly support vertical expansion of linear cities since they preserve space within structures while also providing better efficiencies in multidirectional movement. Instead of expanding our urban footprint by consuming greater green space, we enjoy vertical expansion while preserving the natural environment. From this perspective, the a-ways take up no land at all because they are all used as foundations for linear cities.

Once you consider all of the space efficiencies we gain through autonomous vehicles, a-ways, and Linear Cities, this transportation model allows for a tremendous amount of green space. With such a model, more than half the population in the U.S. alone could reside within the footprint of our interstate system. In states like Iowa, the three highways that cross the state would be more than enough to house that state's population. From this perspective alone, the autonomous transportation system we propose can go a long way to enhancing environmental sustainability.

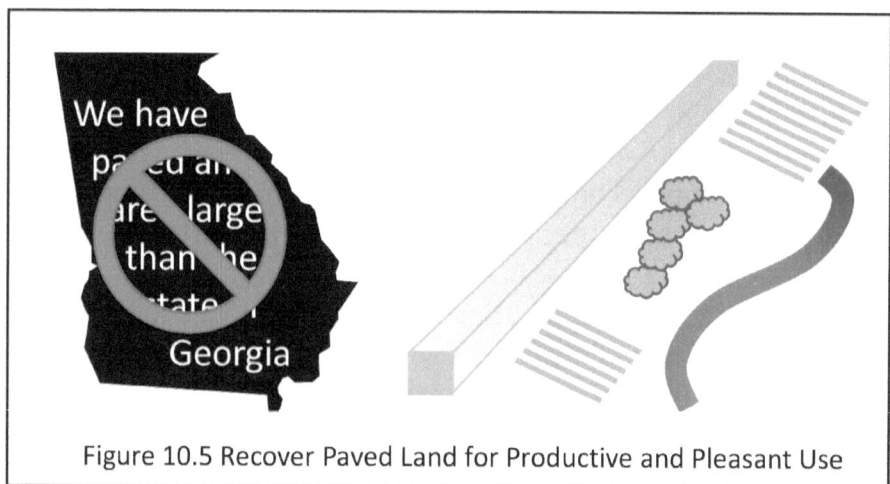

Figure 10.5 Recover Paved Land for Productive and Pleasant Use

Water Management and Sustainability

Water and sustainability are directly related to waste management, infrastructures, and environmental issues. Addressing these areas of sustainability will naturally help in our ability to better manage our water resources. This is important given that water shortages are major problems in many parts of the world today and are quickly becoming global issues. In fact, did you know that 12 percent of households in the U.S. cannot afford public water services today? This percentage is expected to progressively rise as water supplies fall and prices increase. By 2022, this number is projected to be as high as 36 percent of the American population.[81]

According to the American Society of Civil Engineers, over 240,000 water main breaks occur each year in the U.S. This reflects the aging infrastructure systems mentioned previously, and in terms of replacement costs, this would exceed $1 trillion...which we do not have.[82] Water pollution is also an issue. Water runoff from existing roadways results in contamination of rivers, lakes, and oceans, and it also drives up the costs of water treatment for municipalities. Many people are aware of the potential for contaminants in public water supplies, but many choose to solve this through the use of non-reusable water bottles. This again contributes to water and land pollution as well as adding to the 10 percent of municipal waste in the form of plastics.

Some immediate benefits from a-ways can be appreciated in helping these problems. As noted, water infrastructures could be housed within a-ways allowing protection, better maintenance schedules, and greater access for treatments and repairs. The smaller footprint and enclosure of a-ways also reduces runoff concerns substantially, and the absence of oil and gasoline in autonomous vehicles further benefits natural water conservation. But these are not the only ways where our autonomous transportation may help with water sustainability. By using autonomous vehicle delivery systems, "pipe-free" water as well as sewage management becomes an interesting option for the future.

While the specifics of a pipe-free water system will not be covered in this chapter, its introduction in terms of sustainability is worth noting. As described in our examples of autonomous vehicles and medication delivery, the same type of system could be designed for

---

[81] Plos One, "A Burgeoning Crisis? A Nationwide Assessment of the Geography of Water Affordability in the United States." Elizabeth A Mack, Sarah Wrase, January 11, 2017. Retrieved from https://journals.plos.org/plosone/article?id=10.1371/journal.pone.0169488
[82] American Society of Civil Engineers, 2015.

water and fluid waste systems. Special containers could be designed for potable water, beverages, and hot water needs, which would then be delivered to your home just in time for your use. These special containers would be reusable, and once the water was consumed or used, the container would be returned for cleaning and reuse using the a-way system. Likewise, urine and other waste water would be placed in different specialized autonomous vehicle containers from your home. These would then travel in autonomous vehicles to processing sites where the different fluids and wastes could be more efficiently recycled separately into pure water and other useful products, but also into various shades of "grey water" for many uses.

With such a system, we don't have to construct and maintain a completely different water and waste infrastructure. Other than specialized containers and site-specific processing centers, autonomous vehicles and a-ways provide the infrastructure. Likewise, minimal water is wasted since most is recovered and recycled for reuse. And all of this occurs without any significant environmental impact above and beyond what the autonomous transportation system already imposes. We could even collect and use the rain falling on the a-ways rather than adding it to runoff. As you can see, with some innovative thinking and broader applications, this autonomous transportation model provides numerous ways to enhance sustainability of our resources.

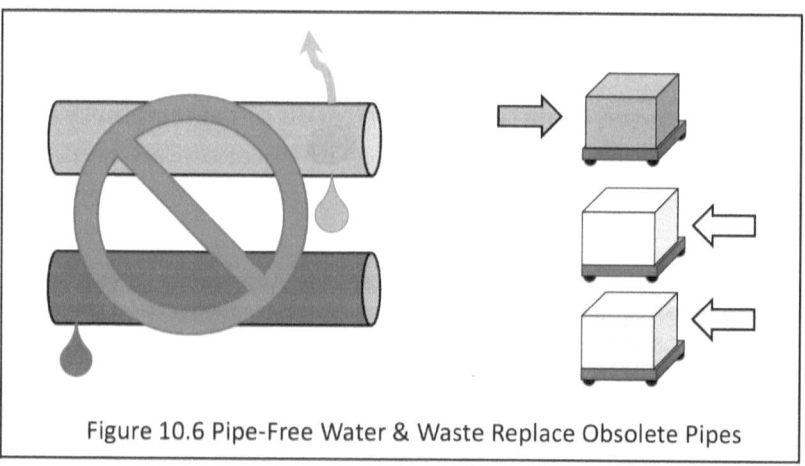

Figure 10.6 Pipe-Free Water & Waste Replace Obsolete Pipes

Not long ago, a serious and lengthy debate was ongoing about the use of an old railway that traveled through the Adirondacks. One group wanted to remove the rails so a bike path could be created. The other group, however, was concerned that the railway might again be used in the future. Their concern was that the railway may return, and the existing infrastructure would be needed. But even if this might happen, the existing rails and roadbed were in such poor shape that a huge amount of money would need to be spent for their repair. Likewise, the chance of rail transportation returning was extremely low. Eventually, the decision to use the rails for bike paths was made. With a-ways a new option becomes available, at any time in the future, an a-way could be constructed along the route with bikes riding on top.

As an outside observer, the decision to turn the railways into bike paths might be a rather obvious one, but change can be difficult. This is especially true when making a paradigm shift from an old technology to a new one. While bike paths are not a new technology, letting go of the old rail system was still difficult for some despite the excessive costs and impracticalities that would result. From the perspective of sustainability, the same could be said about our nation's current transportation system. Several inefficiencies and unsustainable practices exist with today's model, and the time for a significant change is here.

Throughout this section, applications using an innovative autonomous transportation system have been described, and this model greatly improves the efficiency, speed, and convenience of traveling from one place to another. Likewise, the model allows for innovative opportunities to improve delivery systems while addressing privacy and security concerns. While these benefits are the clear drivers of a change that might embrace such an autonomous transportation system, the advantages provided in promoting greater sustainability further support its adoption. As this chapter has demonstrated, the positive impact such a system could have on resource management and the environment is substantial. From these perspectives as well as from the transportation system improvements provided, adopting such a system is not only a rational decision but a responsible one as well. Subsequent chapters provide even more opportunities to enhance sustainability, especially human resources, including time savings, vocations and avocations, health and learning.

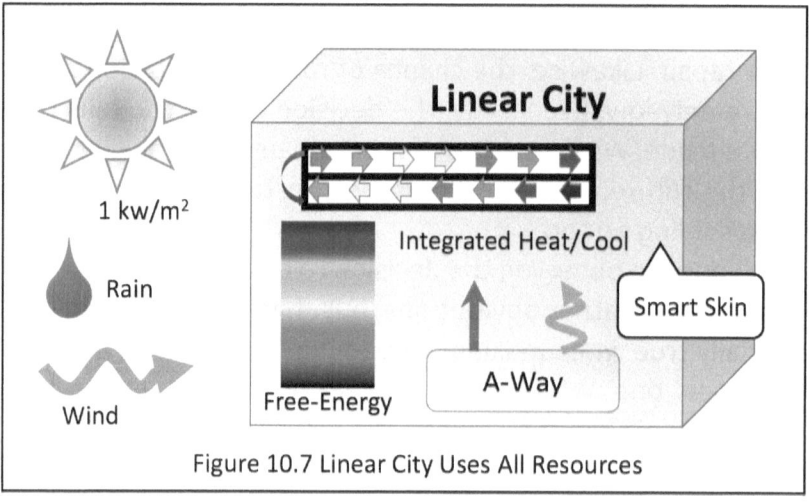

Figure 10.7 Linear City Uses All Resources

11
Challenges from Big Business

THE 2006 DOCUMENTARY *Who Killed the Electric Car*, provided a discriminating look at the forces that squashed the early successes of the modern electric car. Among the most powerful forces were those involving the oil industry, automobile manufacturers, and even the federal government, due to the presumed impact this disruptive technology might have economically on the nation.[83] It is therefore understandable that some concerns should exist regarding obstacles "Big Business" may impose in adopting a comprehensive autonomous transportation system. But whether they oppose it or not, the winds of change are upon them.

In some of the early chapters of this book, historical accounts of various disruptive technologies were described as they related to the transportation industry. The efficiencies of the railroad replaced travel by carriage and horseback. The automobile then challenged rail transportation with the development of new manufacturing techniques and advances in combustion engines. And air travel subsequently played an increasing role in transportation as well. With each of these changes, new technologies came along and forced

---

83 Paine, Chris (Director); Deeter, Jessie (Producer). "Who killed the electric car?" Electric Entertainment, Sony Pictures, 2006.

change. Existing transportation systems didn't necessarily accept these changes gracefully, but they had little choice in the matter when it was all said and done.

By way of comparison, consider the telecommunications industry. Did you know at one point that 99 percent of all homes and businesses in the U.S. had wired phone lines? Today, that figure is less than 50 percent.[84] How did this happen? In a phrase, economic pressure. It was soon realized that a copper loop for a single home cost about $1,000 while a low power digital radio access link cost about $100. Seemingly overnight, telecommunications changed dramatically in the U.S. moving from hard-wired phones to cell phones to smartphones in a matter of a few years. Despite eventually adapting to these new technologies, companies like AT&T resisted the change initially. Likewise, many other telecommunication companies, who failed to adapt, no longer exist.

Notably, the emergence of technological advantages that drove change in the telecommunications industry are also at play in the transportation industry today. Advances have improved opportunities to reduce costs, improve safety and convenience, and provide better speed and efficiencies. However, just as witnessed in the telecommunications sector, resistance to change will occur. In this chapter, some of the more likely pockets of resistance will be described, and at the same time, the futility of this resistance will be highlighted based on our current situation. Several incentives for change exist, and many of these involve things that most big businesses care about (like profits, costs and competitive advantage!). As a result, you will appreciate how these potential obstacles are not likely to be as significant as you might think.

---

[84] Luckerson, Victor. "Landline phones are getting closer to extinction." Time Magazine, 2014. Retrieved from http://time.com/2966515/landline-phones-cell-phones/

Potential Industry Oppositions

If you think about companies that might oppose change in the transportation sector, several immediately come to mind. Automobile manufacturers, gas and oil companies, automotive service centers, and even those businesses that are dependent on current transportation structures and related social behaviors. Let's consider each of these business segments and their reason for opposition before addressing specific reasons why their opposition might ultimately be transformed. Likewise, we can also take a look at how things are already happening within various markets indicating that these changes are truly inevitable.

The most obvious industry to be affected by the changes we propose will be the gas and oil industry. Without vehicles that run on gas and oil, the demand for these fossil fuels will decline quickly. It only seems natural (despite being inherently wrong) to oppose the adoption of an electricity-based transportation system simply from the perspective of corporate revenues and profits. But these companies are already feeling the pinch. In addition to global competition pressures from other oil-producing nations, alternative energy sources that are cleaner and more sustainable are rapidly becoming cheaper than oil. Solar, wind, and battery electricity are already being used by more consumers and businesses than ever before, and governmental and business incentives are being provided to facilitate the adoption of these improved energy sources.

In terms of the automobile industry, their opposition to change should not come as much of a surprise. Remember the resistance of U.S. car companies in the 1970s to pursue more fuel-efficient cars despite advances made by foreign manufacturers in the market? Even today, companies like General Motors and Ford are poorly diversified outside of the automobile sector. This lack of diversification will drive them to oppose any significant change in existing transportation systems because they are heavily invested in the status quo. In other words, they have too many eggs in one basket. But like the oil companies, they are already facing pressure for change as well. In addition to the reduced demand for cars resulting from increases in ride-hailing services, automobile demand is also threatened by self-driving cars. Unless automobile manufacturing companies begin to invest in greater diversification, they will likely end up like the GMs and Chryslers of the world in filing for bankruptcy or being bought out.

Other industries where pushback may occur involve automotive service centers. Electric cars will not require anywhere near the same level of service as gasoline-powered automobiles. And with the autonomous vehicles we propose, service needs will fall even more dramatically. Oil changes will not be needed, braking systems will last much longer, and the number of automotive parts required will be a fraction of those required today. These businesses as well as others will be forced to adapt. For example, autonomous delivery systems could change social traffic at places like Starbucks, and drive-throughs at fast food restaurants like McDonalds. Whether these sectors adapt to the changes, or dig in and resist, remains to be seen. But because of the disruptive nature of the autonomous transportation system we propose, some opposition in these commercial areas can be anticipated.

Why Change Will Win

As discussed, some industries and companies will choose to oppose change, and the reasons for this resistance essentially stem from three motivations. The potential for loss of current revenues, the high cost of investments related to change, and a loss of competitive advantage in the marketplace are the key reasons some businesses will resist. But as we saw from the telecommunications industry, this resistance would prove dangerous. A number of incentives for change already exist, and these will become progressively more powerful in the near future. The market already values companies differently than just their revenues, so Oil & Gas is less important and Automotive does not even make the top 35 companies by market value.

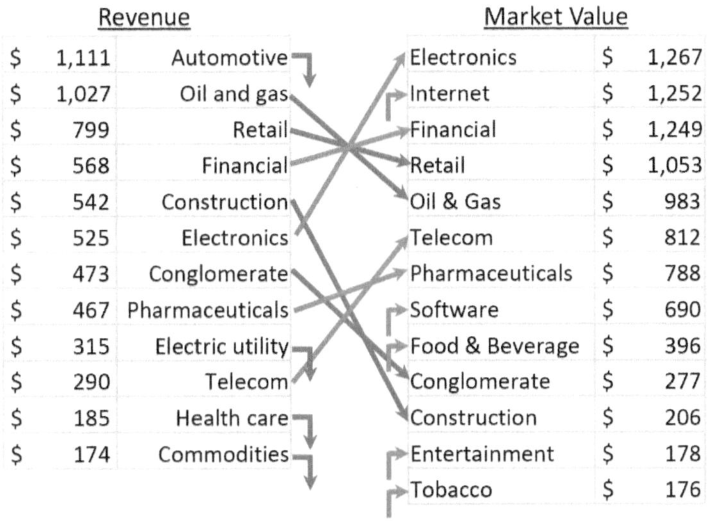

Revenue			Market Value		
$	1,111	Automotive	Electronics	$	1,267
$	1,027	Oil and gas	Internet	$	1,252
$	799	Retail	Financial	$	1,249
$	568	Financial	Retail	$	1,053
$	542	Construction	Oil & Gas	$	983
$	525	Electronics	Telecom	$	812
$	473	Conglomerate	Pharmaceuticals	$	788
$	467	Pharmaceuticals	Software	$	690
$	315	Electric utility	Food & Beverage	$	396
$	290	Telecom	Conglomerate	$	277
$	185	Health care	Construction	$	206
$	174	Commodities	Entertainment	$	178
			Tobacco	$	176

Figure 11.1 35 Largest Companies by 2016 Revenues and 2017 Market Value, $ Billions

Let's take a look at automobile manufacturers first. Tesla offers a great example of how automobile manufacturers will need to adapt to changes as they unfold. Unlike other automobile manufacturers that introduced their electric car as a no-frills, economy vehicle, Tesla chose to create an electric car full of pizzazz and sex appeal. Not only are Tesla models sporty, luxurious, and highly advanced, they are fast! In fact, their most recent model (an SUV) beat a Lamborghini in the quarter mile![85] Who wouldn't want to embrace new vehicle technologies that were not only more energy efficient, autonomous, and safer but at the same time really cool?

---

[85] Ferris, Robert. "Tesla Model X beats Lamborghini and sets new record in drag race." CNBC.com, 2017. Retrieved from https://www.cnbc.com/2017/08/22/tesla-model-x-beats-lamborghini-and-sets-new-record-in-drag-race.html

Tesla showcases other strategies that will also prove to be advantageous through these transitions toward autonomous transportation. Like most other car manufacturers, it is invested in developing self-driving cars, and it is exploring how it can participate within the ride-hailing and car-sharing markets. But more importantly, Tesla is heavily diversified in other innovative technologies including energy. In addition to having its own battery production facility, it is also manufacturing solar roofs and battery walls for homes, businesses, and even electric utilities.[86] These thrusts in clean energy alternatives will naturally be synergistic to their transportation pursuits. Thus, Tesla highlights how competitors in the automobile industries will place pressure on existing manufacturers to either change or be left behind. Elon Musk's stated objective was to change the automobile industry to electric vehicles, and now every major automaker is touting their coming electrics, along with new entrants.

Similar issues exist in relation to the oil and gas industry. In addition to being on the wrong side of climate change and environmental protections, oil and gas companies face pressure from other alternative energy sources. In terms of power generation, solar, wind and battery technologies already provide electricity at lower costs. And in terms of automobiles, higher-end electric cars are selling at comparable prices as gasoline-powered vehicles (actually for less considering the fact they do not require gasoline and oil costs) while offering equal or better performance. We are on the cusp of major change in these areas already. Oil and gas companies must face this fact and determine if they wish to adapt or deal with the market consequences.

Lastly, local businesses will find themselves in the same competitive struggles with businesses that embrace these changes. Just as taxicab companies have struggled to compete with ride-hailing services, companies ready to capitalize on autonomous delivery systems

---

86 Tesla. "Energy." Tesla Website, 2017. Retrieved from https://www.tesla.com/energy

will have market advantages over existing companies that rely on current transportation structures. Tesla provides another good example in this regard as well. Their sales strategy involved selling their cars directly to consumers rather than using dealerships. In some states, dealers collectively fought against this and were able to get legislation passed to prevent Tesla sales in those states. Naturally, this strategy failed because customers simply traveled to the next state to collect their ordered car. Ultimately, the customer always wins when it comes to business markets, and change will occur because customers will demand it. The bottom line is that economics rule, and when new transportation services that are less costly, more efficient, and more attractive become available, the world will adjust accordingly.

Envisioning the Change

At this point, we have discussed why businesses might oppose changes in transportation progress and how some companies (and consumers) are exerting pressure for change in this regard. But being able to envision how this change might happen can be more difficult. After all, those of us who used wired phones and snail mail a few decades ago could have hardly envisioned today's world of telecommunications. So, how will we move from our current transportation system to the autonomous system we are proposing. From a business perspective, how will these changes evolve?

While many creative developments might occur to facilitate these changes, a number of potential scenarios are likely. These scenarios involve commercial-municipality partnerships, private transportation developments, and private residential developments. Let's consider the first possible scenario. Suppose a city is struggling to attract new industries to its region, and they are facing major costs in current transportation system repair. One option for the city would be to begin planning an autonomous transportation system to attract businesses to the area that better met commercial needs. Alternatively, businesses might approach municipalities and request such developments to enhance their operations. For example, FedEx or a university campus might request cities to facilitate autonomous delivery systems using the proposed transportation model. Municipalities would have incentives to pursue these types of changes to attract more residents, more businesses, and greater tax revenues. Businesses would have such incentives to gain a competitive advantage in their markets. These situations are highly likely since they provide win-win solutions to common problems. Several states have already adopted this approach by welcoming self-driving vehicle testing and subsidizing solar energy and electric cars.

Another potential scenario could involve autonomous transportation system adoption within specific sectors where opportunities to reduce costs are essential. For example, the healthcare industry is under significant pressure to reduce costs while improving safety and quality of care. This is particularly true given the aging of the nation's population. Given that autonomous systems could immediately reduce labor costs, enhance medication safety, and improve quality of services, the financial incentives for such industries to invest in autonomous transportation systems are obvious. In fact, the technologies to achieve this are already starting to be available with some healthcare systems exploring such changes.[87]

One last consideration regarding the development of an autonomous transportation model relates to large commercial residential developments. If you have ever visited The Villages in Florida, you quickly appreciate how important transportation is to its residents. With an intricate network of golf cart paths, the more than 150,000 residents can access essentially any location using their golf cart throughout the 32 square-mile campus. This feature, along with many others, have made this seniors-only community the fastest growing

---

87 Muoio, Danielle. "Robots are helping run the first 'fully digital' hospital in North America." Business Insider, 2015. Retrieved from http://www.businessinsider.com/robots-are-helping-run-the-first-fully-digital-hospital-in-north-america-2015-10

area in the nation.[88] If you were a real estate developer, wouldn't this get your attention? What better way to attract residents to your development than providing a fast, fun, autonomous transportation system that met everyone's needs?

The many opportunities offered by autonomous transportation and the other innovations described above form the foundation for start-ups to flourish. Startups create new jobs. As these startups create new products and services, this in turn opens opportunities for new startups. The remaining chapters in this book explore some of the implications of this flowering.

The potential ways our autonomous transportation model might be introduced are numerous, but the ones presented here reflect some of the more obvious ones. Driven by financial incentives and high levels of consumer demand, businesses will become increasingly interested in investing in such systems. At the same time, existing businesses and industries will need to adapt to these changes. It is certainly expected that resistance and opposition from some industries will occur, but the forces of change are too powerful. Ultimately, the market will demand the change, and businesses that embrace it will flourish.

---

88 Rocco, Matthew. "Florida's The Villages is the Fastest-Growing City in America." Fox Business, 2015. Retrieved from http://www.foxbusiness.com/features/2015/03/26/florida-villages-is-fastest-growing-city-in-america.html

12
Job Displacement Concerns

YOU HAVE LIKELY heard fears and concerns expressed by many who believe robots and artificial intelligence will take over our jobs leaving us idle and broke. Did you know that about three-quarters of the nation's population worked in agriculture when the wave of industrialization overtook America? Certainly, most people were then unable to make a living as farmers, but they were hardly idle or broke. In fact, the industrial revolution, driven in part by transportation, created new jobs, millions migrated to urban areas, and the standards of living markedly advanced. Interstate highway construction, air travel jobs, and computer-related opportunities eventually appeared in addition to advances in industries like healthcare, education and entertainment that required human ingenuity and creativity.[89]

There are many competing and conflicting theories on jobs and the economy including: automation kills jobs. Automation yields abundance which helps everyone. The winner-take-all economy enriches the wealthy and impoverishes everyone else. Globalization has hollowed out the middle class. The creative class is thriving in

---

89 Worstall, Tim. "Marc Andreessen On When The Robots Come To Take All Our Jobs." Forbes, 2015. Retrieved from https://www.forbes.com/sites/tim-worstall/2015/08/02/marc-andreessen-on-when-the-robots-come-to-take-all-our-jobs/#40aa768a38a3

the new economy while blue collar and service workers are suffering. Most economists can agree on the desirability of investing in at least two things: infrastructure and education. Infrastructure investment provides jobs and increases future productivity. Education provides people with the skills to lead happy and productive lives. For decades the US was a leader in both these areas, producing an era of improving prosperity for almost everyone. In the last few decades, however, the US has fallen behind the rest of the world in both areas and paid the price. Startups create new jobs. The proposals described above are about a huge new investment in infrastructure, and a future chapter will describe corresponding approaches to improve education.

The auto industry is a good example of this situation. Beset by mergers, takeovers, and bankruptcies, the car companies have already laid-off people by the thousands. Factory workers, dealers, mechanics, salespeople, and others have been victims of these changes. And the remaining jobs are not nearly as good as they once were because bankruptcies allowed the car companies to abrogate their contracts while cutting wages and benefits. However, many other companies, including Apple and Alphabet, are racing to design new autonomous vehicle systems equipped with new sensors and other essential components. Not only have many of these companies become the most valuable enterprises in the world, but they are also vigorously hiring new employees. The rapid growth of the renewable energy industry is creating thousands of new jobs making better use of the skills of laid-off factory workers, mechanics, sales people, and others. And of course, the new kid on the transportation and energy block, Tesla, is growing dramatically.

Economist James Bessen has created a model which predicts that in new industries, improvements in productivity reduce prices and demand increases even more rapidly, so employment increases. As the industry matures, growth in demand doesn't keep up with productivity increases, so employment drops. One key result of this model is that as long as we keep inventing new things that people want, overall employment will increase, despite increasing automation, Thus, as employment in mature industries shrinks, there will be plenty of new jobs. The proposals in this book present just such creative new ideas and can stimulate employment. Subsequent chapters address the training and other needs to permit people to do these new jobs.

In this chapter, we will explore the changes that are likely to occur in the job market with the introduction of an autonomous transportation system. Many of the concerns voiced over such a system in relation to jobs are the same ones heard with automation and robotics in general. On the one hand, we are told that these changes will enhance our productivity and lead to greater wealth. But at the same time, we are also told we are suffering from a slowed economy, stagnant wages, and the threat of declining wealth and that automation will make our situation even worse.[90] Obviously, both scenarios cannot be true, so will greater automation lead to positive or negative changes when it comes to our livelihood? Based on both history and current trends, a very optimistic outlook is supported.

Predicting Job Displacements

New technologies have a way of humbling us when it comes to prediction. Consider Apple, for example. The innovative technology company spent tremendous numbers of hours and ingenuity creating the iPhone, carefully crafting the device's capacity and features. But no one predicted that this device would replace the personal computer the way it has in our daily lives. Retrospectively, it seems rather obvious, but at the time, this was completely unexpected. With that said, predicting how a new transportation technology might change the job market, and our lives, is notably challenging as well.

---

90 Suroweicki, Joseph. "Roboapocalypse not." Wired, 2017. Retrieved from https://www.wired.com/2017/08/robots-will-not-take-your-job/?mbid=nl_81617_p2&CNDID=22832787

In some cases, areas where job displacement will occur can be assumed. For example, traditional car dealers and dealerships will likely be a thing of the past. This has already happened to some extent with customers who buy Tesla products since the company sells directly to consumers online without the inconvenience of haggling over prices. However, other technologies could also make dealerships obsolete. While you might think dealerships would still sell autonomous vehicles to customers, the advances in 3-dimensional printing will probably allow us to "manufacture" many of the items we use at home ourselves. Not only might dealerships be a thing of the past, but many manufacturing tasks will likely move away from traditional industrial plants to our basements or businesses just down the street.

Other transportation jobs can also be assumed to be a thing of the past as well. Once a complete autonomous transportation system is in place, drivers for ride-hailing services will no longer be needed. Perhaps companies like Uber and Lyft might still exist through other services they provide, but human drivers will not be required. Similarly, taxicab drivers will not be needed either, and truck driver positions are already expected to decline in number with the development of autonomous trucks, although there is currently a shortage of truck drivers. Lastly, with the adoption of autonomous vehicles, the need for automotive service workers will fall dramatically. With less service needs and extended periods of time between service, these jobs will certainly decrease substantially. On the other hand, with significantly more autonomous vehicles (small and large) than cars today, the number of people building and working on them may increase. Only time, and ingenuity, will tell.

Other areas of job displacement are more challenging to predict. For example, an autonomous transportation system will not require bus drivers or subway car operators, but at the same time, jobs to help maintain and repair the system will be needed. The number of people using the transportation system would skyrocket in comparison to those using public transportation today, so these types of jobs may explode in number and type. What about jobs related to drive-through businesses? Drive-throughs may not be needed since automated deliveries will be much more efficient, faster, and appealing and because traditional roadways will eventually go away. But these same workers may be needed to prepare food and other items for automated delivery, and they may be able to work for themselves because they can sell directly to consumers. People who like to grow things may find new opportunities, in part due to the rapid growth of higher priced organic foods, but also because they can sell directly rather than earning only a small fraction of the value of their products which currently go to distribution and profits for others. These unknowns make it hard to anticipate which jobs will be completely displaced, which ones may simply evolve, and what new jobs will spring forth.

As with any technological advancement, jobs change. An autonomous transportation system will not be any different. Because transportation affects so many areas of our lives and many related sectors of industry, any major change here will be felt elsewhere. But despite job displacements, this does not mean a loss of job opportunities. In fact, it will likely mean just the opposite. As with previous situations where the adoption of new technologies occurred, the overall number of jobs increased as new types of jobs emerged to replace those displaced. This could be a great time for anyone with entrepreneurial interests, and who doesn't want to be their own boss?

New and Exciting Opportunities

For me, my most creative and productive time tends to be first thing in the morning. In fact, many Americans have the greatest capacity for productivity in the morning because they are more alert and refreshed. In considering this, doesn't it seem a little silly that many of us spend these precious hours sitting in our morning commutes in traffic? Imagine if we could enjoy an autonomous transportation system today. We might still commute to work, but our time would not be spent operating a car, figuring out the directions of the trip, or fuming while stuck in traffic. Instead, we could use this time in a much more creative and productive way.

Richard Florida, author of *Rise of the Creative Class*, sees advances in technology as providing tremendous opportunities in new jobs in a number of areas.[91] As technology has advanced, the trend has been for the actual number of jobs to increase while the use of material resources steadily declines. Think about companies like Apple, Alphabet, and dozens of software application companies. The number of jobs created by these businesses in the marketplace have grown immensely, but the actual amount of non-human resources they use in producing their products is much less when compared to older technologies.[92] It can therefore be anticipated that technological advances in transportation will do the same.

The creative class of workers today comprise about 30 percent of the US workforce, and they earn on average twice as much as individuals in working class jobs and even more than service class jobs. Jobs in the arts and entertainment, technology and media industries have grown tremendously in number while offering higher wages and a better quality of life.[93] These shifts are already occurring as a result of technological changes in other industries. With the Internet and social media as well as advances in communication services, new companies and jobs have emerged in software application development, online commerce, entertainment and many other areas. And we are still just scratching the surface on the potential of these new technologies.

---

[91] Florida, Richard. The Rise of the Creative Class--Revisited: Revised and Expanded. Basic Books (AZ), 2014.
[92] Ibid.
[93] Ibid.

If the effects of new technologies on other sectors provide some insights, we can anticipate we will also enjoy other quality of life benefits with changes in transportation. People who today are excluded from many jobs because they cannot get around will be able to choose occupations that interest them and use their skills. For one, autonomous transportation will likely allow us to have more time for recreation and hobbies. From an employment perspective, this means specific industries related to health, arts, and entertainment will have increased consumer demand to drive new ideas and new jobs. Likewise, with autonomous delivery systems, an array of new service job opportunities will emerge that bring unique items to people's homes and businesses. And design jobs will likely flourish in many capacities as well…perhaps eclectic autonomous vehicle design packages might be developed so you can "3-D" print your own recyclable containers and even autonomous vehicle at home! How cool would that be?

With autonomous transportation systems, job displacement will occur, but at the same time, many new and exciting jobs will be created. Not only will these new opportunities for employment be more creative and personally fulfilling, but they will also provide an overall increase in our collective quality of life. The standard of living will improve for everyone as a greater amount of time will be spent away from things like traffic, concentrating on the road, and taking care of and paying for our vehicles. And concurrently, many new services and offerings will arise as our jobs will allow the use of our creative energies to a much greater extent.

US Employees (thousand)		35 Highest Revenue Cos. (thousand)	
Professional, business services	20,136	Retail	2,845
State, local government	19,428	Construction	2,138
Health care, social assistance	19,056	Automotive	1,912
Retail trade	15,820	Electronics	1,166
Leisure, hospitality	15,620	Oil and gas	1,052
Manufacturing	12,348	Financial	1,002
Nonagriculture self-employed	8,733	Electric Utility	926
Financial activities	8,285	Conglomerate	911
Construction	6,711	Telecom	429
Other services	6,409	Health care	230
Wholesale trade	5,867	Pharmaceuticals	120
Transportation, warehousing	4,989	Commodities	93
Educational services	3,560	**TOTAL**	**12,825**
Federal government	2,795		
Information	2,772		
Agriculture wage, salary	1,501		
Agriculture self-employed	850		
Mining	626		
Utilities	556		
TOTAL	**156,064**		

Figure 12.1 Employment by US Sector and Global 35 Highest Revenue Companies, 2016

How to Get There from Here

When major industrial factories were being constructed in cities, and when technologies began to threaten farmers' way of life, we can imagine they had many concerns about how they would acquire new skills needed for the jobs that were emerging. The same concerns exist today. Unfortunately, our country does not have a systematic way to educate and train adult populations, and this would be a tremendous asset if this were in place. Community colleges offer some support in this regard, but many skills and abilities required for emerging industries are not necessarily easy to find for adults. In addition, a certain degree of ambivalence often exists among adults about pursuing additional education and training at times due to a lack of personal or professional incentives. And the development of new skills can be undermined by a degree of social inertia (or even resistance) especially when seniors are considered.

Though educational and retraining issues exist currently, this does not mean incredible opportunities are not present in overcoming these challenges. Let me provide a perfect example. One of my friend's twin children were in their junior year of high school, and they both wanted to "pad" their application for college enrollment. As a result, they thought about various community service projects they might do. In an "ah-ha" moment, they came up with the idea to create a non-profit organization to help seniors learn to use new technologies like iPhones, laptops, etc. With a little bit of planning and support from their high school and friends, they created "Savvy Seniors." Much to their amazement, enrollment went quickly from a handful of seniors to more than two dozen. This is a perfect example of the new and creative opportunities that will exist as we transition from the old to the new, and the opportunities for seniors and youth to work together.

This is not the only example of creative ways of filling the gap between new job requirements and the skills needed. Through online programs, many curriculums now exist for adults making it easier and more affordable to receive training and education in specific areas. As the demand for different training and education increases, private industries and even individuals will likely seize these opportunities by providing additional programs. A structured system or business that enabled adolescents and college students the chance to earn money by providing these services to adults may also evolve in the future as the job market changes. Which option or options may turn out to be the best is yet to be determined, but the capacity to have education and training systems in place for adults and adolescents alike are significant. And a better transportation system will make it easier for people to get to these new training opportunities.

With this in mind, the path to get individuals "up to speed" for new jobs and employment opportunities will probably involve a combination of business, social, and political changes. At the same time, technologies will continue to provide new prospects for entrepreneurial endeavors in a number of areas. The adoption of autonomous transportation systems should not be viewed with fear, worry, and concern simply because things will change, and some jobs will be displaced. This has occurred repeatedly with dozens of technological advancements over time. Instead, we should embrace these changes with an understanding that new and exciting jobs will appear that will promote a better standard of living for us all. The chapter on learning will provide more specific innovations to assist in this important aspect of society.

13
Social Challenges

AMERICAN SOCIETY AND culture stand for a number of things. According to John Truslow Adams, "The American Dream is that dream of a land in which life should be better and richer and fuller for everyone, with opportunity for each according to ability or achievement."[94] Individualism, a hard work ethic, and creative ingenuity are among some of the more recognizable characteristics we have come to associate with these American ideals. With this in mind, nothing depicts American freedoms quite as well as a 16-year-old getting their driver's license for the first time and getting behind the wheel. A sudden sense of freedom, maturity, and autonomy comes over them as they pull out of the driveway with unbridled excitement. For many of us, that was one of our most precious memories and an inherent rite of passage.

The automobile as we know it today has been a part of our society for over a century. It has outlasted the radio, the phonograph, and many other innovations despite many important advances in technology that could have made cars a thing of the past. Why is this so? We have already talked about some of the reasons…the influence of big business, a thriving industry that supports millions of workers, and a natural resistance to change. But other reasons exist as well, and one

---

94 Adams, James Truslow. *The epic of America*. Transaction Publishers, 2012.

of the most noteworthy relates to our culture and society. Freedom, individual expression, the value of ownership, and even nostalgia reflect social values linked to the persistence of the classic car.

Social challenges related to these values still exist today, but they may not be as significant as you think. American culture and values continue to reflect the same as those just described, but generational lifestyles and trends are shifting, and these offer key insights about how our car culture is changing as well. These also highlight how autonomous transportation will not only align well with these trends but also continue to meet our core values as a society. But before we begin discussing these aspects, let's first begin with the specific social and cultural challenges an autonomous transportation system might face in America today.

Identifying Specific Social Challenges

According to the U.S. Bureau of Transportation, each household in the U.S. has 1.9 vehicles. What is more interesting is that the average number of people per household is only 1.8.[95] One significant reason for this occurrence is our need to be "ourselves." As an individualistic society, we thrive on self-expression, creativity, uniqueness, and looking out for number one. Cars have naturally fit into this concept for decades because they allow us to not only be ourselves but to have our own private space. In other words, cars let us be the individuals we want to be while allowing us to showcase this individuality to everyone as we drive along.

Individual expression can occur in many ways. Some may choose to have purple hair while others adopt a more "goth" appearance. Others may be more traditional in style but select the right colors and clothing designs to make their own personal statement. Cars are simply an extension of this expression, and car manufacturers have

---

[95] U.S. Bureau of Transportation. "Highlights of the 2001 National Household Travel Survey." Website, 2001. Retrieved from https://www.rita.dot.gov/bts/sites/rita.dot.gov.bts/files/publications/highlights_of_the_2001_national_household_travel_survey/html/section_01.html

certainly taken advantage of this fact. We have all seen the stylish, sports sedan weaving its way along a Pacific coastal highway giving us a glimpse of the freedom and luxury we might be able to enjoy. While the entire advertising industry spending in the U.S. totaled $179 billion in 2016, major automakers accounted for $33 billion of this figure. Despite automobile manufacturers accounting for about 3.5 percent of our nation's gross domestic product, they account for about 18 percent of advertising revenue.[96] It would appear the car manufacturers have a pretty good appreciation of our need for self-expression based on these investments, and over the years, they have helped mold our view of the American Dream.

Automobiles can also reflect other social values than self-expression and fashion. Think about their relationship to socioeconomic status. Owning a luxury vehicle is like wearing a stylish, Italian-designed suit or owning an estate property on a lake. Knowing the price tag that goes with a specific vehicle gives others the impression you are doing pretty well (even though all your credit cards might be maxed out!). Cars can also reflect lifestyle needs. The family of five that drives a minivan comes to mind as does the outdoor enthusiast who loads his SUV with canoes and the like (yours truly). While this latter feature can be a more functional trait of a car, it also can portray a particular image as well. For example, most people don't really need a Hummer for their everyday travel on suburban streets, yet some choose to adopt this "image" as part of their persona.

Lastly, we come to the car enthusiast, or as some may prefer, the car fanatic. Though these individuals may actually drive their precious possessions very little, the enjoyment they receive in owning such a vehicle is tremendous. Perhaps nostalgia fuels this passion, or maybe they simply have a love for how the automobile was built and how it performs. Regardless, the value cars have for these individuals

---

[96] Statistica. "Largest automotive advertisers worldwide in 2015 (in billion U.S. dollars)." Website, 2016. Retrieved from https://www.statista.com/statistics/671530/automotive-advertisers/

is both different and substantial, and it also plays a notable role in American heritage and culture. Likewise, there is something about our culture that values ownership in general. Being in possession of a vintage car or luxury vehicle offers some level of satisfaction and security despite the fact that buying a new car tends to be a rather poor investment. This speaks to the power of social trends.

Addressing These Challenges through Autonomous Transportation

At first glance, you might be unable to see how autonomous transportation might accommodate our cultural needs based on the role that traditional cars have played in our history. But as already mentioned, new car companies like Tesla have taken these social values into account in their approach to new car design and sales. Despite being electric, Tesla models are stylish, luxurious, and represent a status symbol for many. Likewise, various design features allow customization and individualization. Now, simply take this a step further and apply these same concepts to an autonomous vehicle or personal mobility device. Just because a vehicle is autonomous doesn't mean it can't be unique, stylish, and expressive. And because they are not burdened with excessive weight or size, and because they are much more affordable, they can be much more easily customized based on individual preferences.

As you recall, autonomous vehicles are divided into mobility platforms and containers. Since the mobility platforms house all the driving and control functionalities, the container can in essence be designed in countless ways. You may choose to have a basic container to get you from one place to another, or you might "jazz it up" with different colors, shapes, lights or a plush interior. You might customize it yourself and be truly individualistic, or you might purchase a design from a number of businesses — auto customization has long been a strong neighborhood business. As long as the containers meet size dimensions and have a secure standard interface with its mobility platform, they can be uniquely designed any way you like.

While some autonomous vehicles might simply have the basic features to accommodate autonomous travel, others might be upgraded to include a number of features a particular person may want in addition to unique appearances and designs. Special features could be created within containers to accommodate individual or specific transportation needs. For example, autonomous vehicles can be designed with interiors for the safe transport of children, recreational equipment, refrigerated items, and many other contents.

Options will also exist in relation to autonomous vehicle ownership. For some, possessing their own autonomous vehicle may seem ideal, especially if they choose to customize their autonomous vehicle to their own liking. But at the same time, options to share autonomous vehicles will be available. Just as ride-hailing car services currently allow such "sharing" of cars, autonomous vehicles could be shared among many users at different times allowing greater freedoms and cost savings. Some may own a customized container but used shared mobility platforms. In fact, some may choose to own their own customized autonomous vehicle for their own personal use, but also rent it out to others when they do not need it. "Themed" autonomous vehicles might even represent a business opportunity for some entrepreneurs for specific events and occasions, think Star Trek or Star Wars.

Despite the presence of autonomous vehicles and an autonomous transportation system, there would still be room for nostalgic car collectors as well. In fact, statistics highlight the fact that car enthusiasts actually drive their vintage automobiles very few miles each year.[97] Just as with other nostalgic items (like vinyl records and comic books), different market segments would be developed to accommodate these interests and hobbies. There is no denying that the classic automobile will always have a place in America's heritage,

---

[97] Birch, Stuart. "Classic car enthusiasts profiled." The Telegraph, 2010. Retrieved from http://www.telegraph.co.uk/motoring/classiccars/7398696/Classic-car-enthusiasts-profiled.html

and opportunities will certainly exist for car enthusiasts to pursue their passions in this regard.

While many features of the autonomous transportation system proposed require some standardization to enjoy the efficiencies and benefits described, the system is also adequately flexible to accommodate social values and cultural preferences. If anything, autonomous vehicles would facilitate personal expression to an even greater extent while enhancing other opportunities for creative innovation in design, shared services, and various business endeavors. And all the while, we would enjoy a much more efficient and effective transportation system using fewer resources.

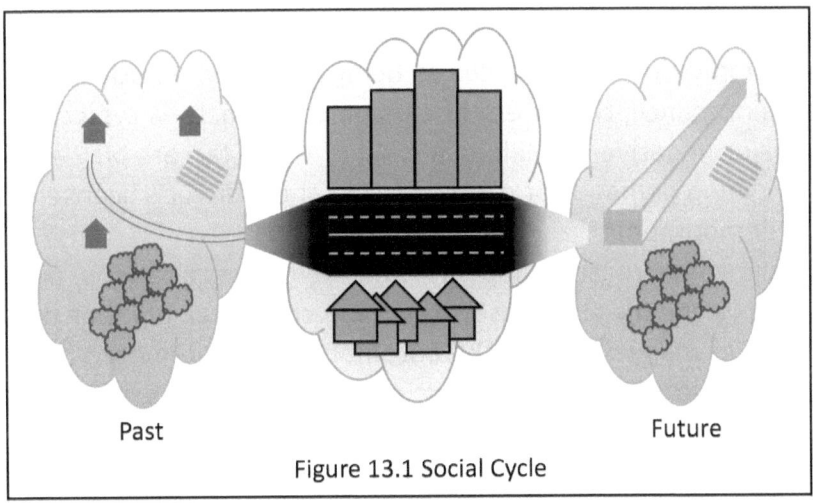

Figure 13.1 Social Cycle

Shifts in Perspectives and Philosophies

While autonomous transportation can accommodate many of the traditional social and cultural preferences associated with the classic automobile, it should also be noted that this is not likely to be necessary in many ways. For one, economic drivers will continue to take the lead when it comes to change. At the same time, however, social pressures for change also exist. We started this chapter discussing the typical experience that a teen might go through in getting a driver's license, but this rite of passage is fading quickly. In 1983, 46 percent of 16-year-olds had their driver's license. However, by 2010, this figure had fallen to 28 percent![98] Regardless of the reasons for this decline, the pressure to get your license to drive as a teen is less, and this has social implications when it comes to transportation.

Let's consider the reasons for this declining demand for teen driver licenses. First, cars are not cheap. Even if a teen had a license, the ability to afford a car is more challenging than in the past. This is hindered by a limited job market for this population segment as well. In addition, many other less costly (and less demanding) options exist. Ride-hailing services offer great solutions for teens without a license. Likewise, many families have moved into urban areas where public transportation systems are available for teens. And many teens simply do not see a car as being "cool." In addition to costs and maintenance, the effect of cars on the climate is typically seen as a major negative by Millennials. All of these factors play a role in the declining demand for car ownership and a driver's license in this social demographic.[99]

While these reasons appear to be logical explanations, another interesting theory also exists regarding the lower number of teen drivers. For many decades, the automobile was the means by which extracurricular social interactions occurred. For teens, the opportunity to interact with other teens outside of school functions in the past were limited unless you had a car. But that is no longer the case. Teens have the Internet and social media to meet these social needs, and they are less concerned about their friends and acquaintances being in the same geographic area. As a result, cars are less beneficial than they were previously from a social point of view.

---

98 Samuelson, Robert J. "Is the car culture dying." The Washington Post, 2016. Retrieved from https://www.washingtonpost.com/opinions/is-the-car-culture-dying/2016/07/10/52a20a56-451e-11e6-88d0-6adee48be8bc_story.html?utm_term=.8860596a877e

99 Ibid.

This tendency also highlights another important social trend… mobility. We can all appreciate the portability and mobility of our digital devices, and increasingly, the world has become much smaller with globalization. Individuals as well as families relocate often in today's world whether it is due to preference, career, or other motivations. What does this have to do with cars? For some, owning a car means having to register the vehicle with every new relocation. For others, they may use their car too infrequently because of their out-of-town travels or because of limited urban parking. Demands for mobility have therefore encouraged less ownership and greater sharing and ride-hailing activities. Here again, we are seeing a social shift in perspective and philosophy that is having a direct impact on transportation choices.

Many of these shifts in perspective are generational in nature. Millennials see the world differently than Baby Boomers, and as a result, their transportation needs are different. Because they have different values, different social behaviors, and their own unique culture, they will have a major impact on transportation system demands and ultimately the systems in place. We can only wait to see how Gen-Z characteristics will differ from Millennials. In many ways, the autonomous transportation system being proposed aligns well with the shifting philosophies currently occurring. Given these changes and given the capacity of the system to meet individual needs, the actual social challenges posed are likely to be less significant than expected.

14
Security and Privacy Protections

SECURITY AND PRIVACY protections have become one of the most important issues today in our country if not throughout the world. Many major corporations and institutions have been victims of database breaches resulting from illegal hacking activities, and the number of identity theft crimes continue to increase each year. In addition, media stories periodically reveal how domestic or foreign governments have "tapped" into our personal information or digital behaviors for a variety of reasons. And if that isn't bad enough, we have to remember logins and passwords for a plethora of websites and accounts, with each having different symbol, letter and capitalization requirements!

Unfortunately, a focus on security and privacy protections is not about to go away, and these concerns will increase in the transportation sector as well. Why? For the simple reason that any new system will have more robust computing power equipped with new software applications and abilities. This is already the case if we look at existing car companies today. For the most part, older automobile manufacturers considered security protections as an afterthought, and several episodes of hacking have been reported. Keyed locking systems have advanced to fingerprint, smart keys, smartphones, and digital keypads, but few revolutionary changes have occurred. In contrast, Tesla vehicles have been described as computers that happen to have a car attached to them. Not only are software applications an essential element of Tesla vehicles, they are unique in allowing remote downloads for system upgrades, and Tesla is highly committed to system security.

From a practical perspective, you can appreciate that security and privacy issues become more important as additional software is added to any system. Information and data are highly coveted items. To highlight this point, the total number of property crimes in the U.S. is around 8 million a year with less than 9 percent involving motor vehicles.[100] But the number of crimes involving identity theft approaches 18 million a year![101] Now consider an autonomous transportation system. Not only is such a system highly invested in software and computing power, but it also contains much more than just your identity…it contains you, your goods and even your children! As a result, security and privacy protections are extremely important, and any autonomous transportation system design must address these protections at every level to be desirable.

In many instances, security systems fail due to a lack of foresight and planning. Therefore, we would like to address key areas where security and privacy protections will be needed in an autonomous transportation system, and present effective solution approaches in each of these areas. From physical dangers to cybersecurity threats, these concerns will be considered in a comprehensive way due to their

---

100 Federal Bureau of Investigation Uniform Crime Report. "2015 Crime in the United States." FBI Website, 2016. Retrieved from https://ucr.fbi.gov/crime-in-the-u.s/2015/crime-in-the-u.s.-2015/offenses-known-to-law-enforcement/property-crime

101 Bureau of Justice Statistics. "17.6 million U.S. residents experienced identity theft in 2014." BJS Website, 2015. Retrieved from https://www.bjs.gov/content/pub/press/vit14pr.cfm

heightened importance in today's world. Likewise, these solutions will demonstrate how such concerns can be effectively resolved to permit safe and secure transportation for us all. In this circumstance, an ounce of prevention is certainly worth a pound of cure, or more.

A-Way Security and Safety Protections

Throughout the book, many of the hazards related to the physical environment have been discussed in relation to cars and trucks. Deer, potholes, and inclement weather are a few of the major threats to transportation safety that exist today. And a-ways eliminate these hazards simply by nature of their enclosed structures. But these same enclosures offer many additional advantages when it comes to security and safety. A-ways can block electronic signals from passing in or out, thus eliminating many hacking opportunities. One of the most important relates to opportunities for continuous transportation system monitoring. Each component of the autonomous transportation system would test both itself and its environment to detect anomalies and malfunctions. By having comprehensive monitoring in place, options for a number of response systems could be employed to enhance transportation security.

With a robust monitoring system, each autonomous vehicle could be effectively tracked within the a-way enclosures. Any sudden malfunction would alert the system that something needed to be addressed, and any autonomous vehicle in distress could be attended to immediately. There would be no more waiting on the side of the road for an hour for a tow truck. Instead, many problems can be addressed remotely, and the system could dispatch necessary vehicles, repair personnel, or relocation specialists to immediately address the issue. Such a system would thus detect any anomaly and be able to take corrective action so that transportation was not affected, and so passenger safety was preserved.

While each autonomous vehicle would be tracked and monitored within the system with some broad security measures in place, the contents of the autonomous vehicles would not be something the system would necessarily "know." In this way, privacy protections would be preserved (regardless whether contents were packages or people). At the same time, reusable containers would be monitored for unauthorized tampering or access. In fact, many cargo containers implement such a system now. Through radio frequency identification (RFID), cargo container movement and progress are tracked, and electronic seals are in place to allow detection of container tampering or unwanted access. A-ways would simply provide a more comprehensive environment in which these security and privacy measures could be employed.

A-ways could also provide oversight for a variety of emergency situations. If a vehicle malfunction should occur, or even if an autonomous vehicle was purposefully altered in some way, a management override function could allow autonomous vehicles to be redirected or a-way lanes to be closed and avoided. Emergency response vehicles could be dispatched within seconds if needed. The system could even impose physical constraints (through other autonomous vehicles or internal a-way structures) on a specific autonomous vehicle's movement or a-way lane access if security and safety protections were required, for example preventing a thief from getting away. All of these measures could greatly enhance the security and safety of our transportation simply by having an enclosed structure in which better surveillance, monitoring, and control can be provided.

In addition to these features, any security system that allowed management overrides would naturally need to have enhanced security itself to ensure only authorized access could occur. Protections against hackers into the system would need to be robust and highly developed, no different than other major digital database systems today. But with these measures in place, the ability for a-ways to provide enhanced physical security and safety measures can be readily appreciated. This capacity alone would be a tremendous step forward when compared to our current transportation system.

Autonomous Deliveries, Security, and Privacy

While security and safety are notably improved within a-ways, concerns regarding both security and privacy will exist in relation to autonomous deliveries. As packages and items travel autonomously from sender to receiver, the autonomous transportation system must be able to have safeguards in place. For example, senders do not need to know specifics about the recipient since all navigation and controls will be managed by the autonomous system. Likewise, the actual contents of the sealed autonomous delivery containers do not need to be known once they leave the sender unless a threat to public safety exists. In this way, privacy can be better ensured.

Consider our delivery systems now. Whether it is through the U.S. Post Office or through courier and freight companies, senders provide the destination address, the name of the recipient, return address, contact telephone number, and other important pieces of delivery information. Likewise, numerous handlers along the way may be able to recognize this and general aspects of a package's contents. Do all these people really need to know you are receiving a medication package from your cardiologist? Wouldn't it be better if all of this information was private and protected? An ideal system would be able to accomplish this while complying with key privacy regulations like HIPAA. This is what an autonomous transportation system can actually provide.

The autonomous transportation system has three major applications controlling the details of the delivery process. The first major application is the Delivery Negotiation application. Today the recipient has little control over the timing of a delivery, other than how much of a premium to pay for expedited delivery. The recipient guesses how long shipping will take, places the order, and gives the sender a delivery address. With the Delivery Negotiation application, the recipient is in control of exactly when and where a delivery takes place. The sender doesn't have access to this private information. This data is personal, so it is kept private and secure by the system. In other words, the sender would have little if any knowledge about the recipient's address, delivery times and dates, or other information. The sender would simply use the Delivery Negotiation application to gain approval to send the item, and when to enter it into the autonomous transportation system.

The second major application is the Routing application. The Routing application is responsible for finding an optimal path for each autonomous delivery. This is somewhat similar to a map routing application, but much more powerful. The autonomous routing application has much more information about the current situation, the planned future usage of each section, and historical statistics. However, the information about the specific route being taken is not shared outside the part of the application dedicated to your autonomous vehicle's path, so again your data is private.

The third major application of the autonomous transportation system is the Navigation application. The purpose of the Navigation application is guiding the moment-by-moment action of an autonomous vehicle along its path. Thus, it has limited information beyond the current cloudlets. The Navigation application interacts with the Routing application to decide when to move from one segment to the next but has no overall information about the end-to-end route. Of course, the final phase of the navigation is delivery to the specified address, but all information is secured as soon as the delivery is made because the Delivery Negotiation application securely manages this information.

Let's go through a typical scenario using the autonomous transportation system for a medication delivery. Your pharmacy has a medication that needs to be delivered to you, so you can take your treatment as prescribed. The pharmacy (the sender) issues a delivery request message, which identifies the delivery type, specific service needs, the recipient, payment methods, and other relevant information, but not the recipient's address. The Delivery Negotiation application sends a delivery approval request to you (the recipient). At this point, you can either approve or reject the request. If you approve, the application then works with you to determine the specifics about your delivery preferences, with the Routing application. The system dispatches an autonomous vehicle to collect and deliver your medication, and then updates you and the pharmacy with select delivery details. The Navigation application then performs segment by segment navigation within each cloudlet.

One of the benefits of such a system is its ability to send you packages and items to any location you wish. For example, suppose you are going to be on a ski vacation next week. You could arrange a different delivery date through the system, but you could also direct the delivery to wherever you might be at the time. Suppose you need the medication right away. You would simply provide the autonomous delivery system with the hotel's address and your room number, so the item could be delivered to you at the time you needed them. And because this would be negotiated through the delivery system, no one else would know your whereabouts or the fact that you were not at home.

Because the delivery is managed by the autonomous transportation system, only select information is provided to both the sender and the recipient. This offers the ability to maintain data privacy and security at a system level. Likewise, because a delivery cannot be arranged without an approval certificate by the recipient, an unwanted item cannot arrive unexpectedly to your home or office. And despite arranging the specifics of the delivery, the system would not be aware of the contents of the delivery container or have access to any personal information about you other than that needed for delivery. Not only is this type of delivery system much more secure and safe than today's systems, but it also enhances personal privacy protections significantly.

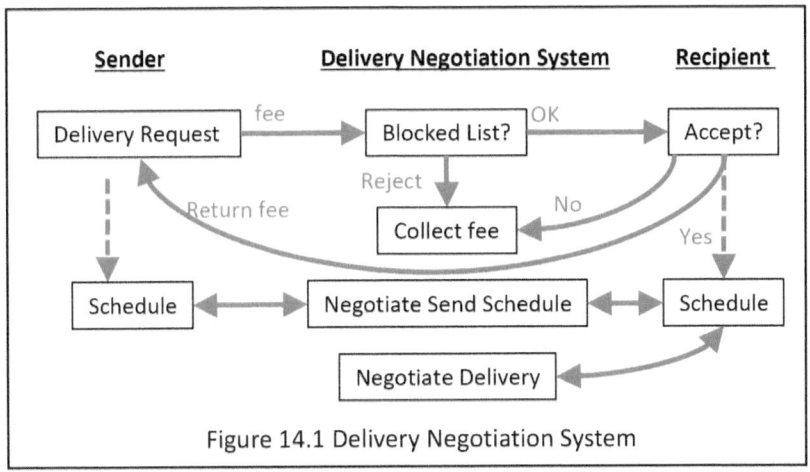

Figure 14.1 Delivery Negotiation System

Deliveries and Securing Your Premises

A natural concern with any delivery system pertains to the security and protections related to your home, office, or other location. Regulating which packages and deliveries are permitted to enter your premises is essential, and this even extends to visitors, repair persons, and installers. With this in mind, the presence of an autonomous door would provide added security and privacy protections for an autonomous delivery system. As autonomous delivery vehicles arrive at your home (after recipient approval and a delivery certificate was validated), these would interact with your autonomous door even in your absence to complete the delivery. The same system would be used for autonomous vehicles carrying individuals who may need to access your home or office.

What exactly is an autonomous door? In essence, an autonomous door is a collection of hardware, software, and protocols to control entry into your premises. In terms of hardware, an autonomous door is a physical door that can be opened under system control along with one or more holding areas. The holding areas allow items to be delivered without entering farther into your premises, and securely stored until you are ready to do something with the items. In addition, different holding areas could accommodate different container needs. For example, you would likely have a refrigerated holding area for the delivery of groceries. Note Amazon has recently proposed a more limited version of this autonomous door, but without the holding areas.

While this system is one logical consideration to meet security and privacy needs, the system is also flexible for other considerations. For example, you may be comfortable with a particular autonomous delivery vehicle entering your autonomous door and placing your item in a specific location in your home or office without going to a holding area. This might be advantageous for just-in-time medication deliveries where the medication (along with a glass of water) could be delivered right to you in your home. It also could be ideal for appliance or furniture deliveries. In these instances, the autonomous door might permit access of the autonomous vehicles carrying your appliance and an installation technician to your premises. And it would also regulate and monitor the removal of the old appliance (to ensure nothing else was removed from your premises besides this item).

As with other parts of the autonomous transportation system, autonomous doors would also have the ability to identify autonomous vehicle containers, individuals, and specific items arriving at your home or office to help ensure security and safety. Autonomous doors provide you the option of remotely viewing deliveries, communicating with people and applications, and directing actions. Autonomous doors would also likely have the ability to monitor and provide continuous surveillance of your premises. By having this added interface between autonomous vehicles arriving to your premises and your actual home or office, you control access as well as how you would like deliveries and arrivals to be handled. As a result, autonomous doors would markedly enhance security while also providing secure flexibility and convenience.

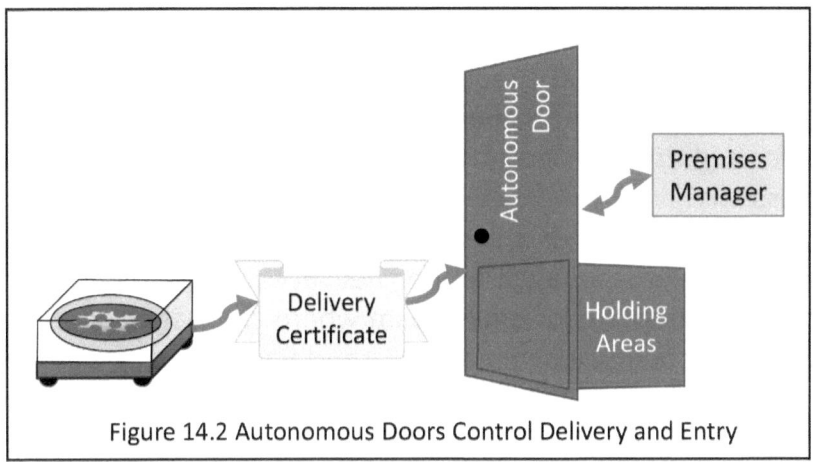

Figure 14.2 Autonomous Doors Control Delivery and Entry

A Word About Spam and Public Security

We are all victims of junk mail and spam. Whether it arrives in our mailboxes, our email inbox or junk folders, our phone and voicemails, or even through text messaging, we are all quite familiar with this albatross. Having explained how an autonomous delivery system might work, you might be concerned that such a system offers one more way where spam could enter our lives. The delivery management process keeps spammers from sending you physical objects without your prior approval. What if a sender kept sending delivery requests to you through the system despite repeated disapprovals? Not only would that be quite annoying, but it also could reduce the system's efficiency in dealing with these "junk" requests.

In actuality, there is a simple and effective way to handle this situation, and in fact, such a system could actually be employed today to limit spam email, telephone calls, and junk mail. Because autonomous deliveries are managed by the autonomous transportation system, the system serves as a regulator for spam and legitimate delivery requests. One very effective way to manage spam would therefore be to charge senders a fee for every delivery request entered. For requests that were approved, the fee would be returned to the sender by the system. But for each disapproval, the fee would be kept. And if repeated disapprovals were received from the same sender to the same recipient, fees could actually increase to offer additional disincentives against repeated delivery request submissions.

In considering such a system, the fee would not need to be substantial. Perhaps, even $1 per delivery request might suffice. Unlike mass mailings or email spam where thousands of contacts could be reached at a minimal cost, the same approach would cost thousands of dollars through the autonomous delivery system. In addition, you could maintain an approved list of senders and an unapproved list of senders for deliveries in the system. Once on the unapproved list, you would not even receive delivery requests from that sender any longer. Only approved senders would be able to forward requests to you for delivery if you wished. These offer some very simple, yet effective, solutions to regulate spam within such a system.

In addition to spam concerns, unfortunately we have to worry about larger and more serious issues as well. You only need to walk through the security checkpoints at the airport to appreciate these larger concerns. But at the same time, the hassles and inconveniences associated with these security checks are significant. We have to wait in long lines to eventually partially undress, partially unload, and be body scanned before redressing and repacking to proceed to our gates. And recent events highlight the concerns over the risks associated with using vehicles as weapons. While we seek to address these issues with the current transportation system, it is clear these protections are limited and associated with significant delays.

In this regard, the autonomous transportation system again offers significant advantages. From the standpoint of autonomous package deliveries, the system would require senders to have registered accounts in order to send their items, and likewise, recipients would also have to be registered within the system to receive specific types items at the location. Specific identification requirements would be needed in order for the delivery transaction to occur, and this would offer some transparency and security in the process. And because the recipient has accepted delivery, packages can't be sent to non-existent or unsuspecting addresses. At the same time, while exact contents of the packages would not be known, in order to preserve privacy protections, scanning for potentially harmful substances could occur. With this in mind, flags might be raised for specific size, weight or composition parameters, or specific senders or recipients. And if flagged, a container on an autonomous vehicle could be sent en route for screening in more detail. This ability for en route testing has other uses, for example, you could test the salad you ordered for E. coli and other pathogens, or that the tomato is ripe and of the type you ordered.

These same types of screenings could be performed for autonomous vehicles carrying individuals as well. In essence, many of the same procedures used today at airports and buildings would likely be used. However, there would be one major difference. The autonomous transportation system would permit these processes to be performed without interfering with our actual transportation. For personal scanning, the autonomous vehicle would just move you through the scanners while still in motion. For package deliveries, the system would perform these functions while the delivery was in route, and while still adhering to delivery schedules. And in terms of a-ways, screenings would be incorporated into routine travel schedules. Only people and items identified as requiring additional screening might be delayed, and even these delays would be much less in duration compared to today's transportation and delivery systems. As you can see, societal security would remain a concern, but the autonomous transportation model described would offer much better solutions to these issues.

With the adoption of digital technologies, software applications, and autonomous systems, data security, personal and public safety, and privacy protections become higher priorities. This is true of any system, but it is particularly relevant to transportation since much of the time the cargo will be quite "precious." At the same time, such a system will require attention to how transportation vehicles interact with our homes and offices. These represent additional security and privacy areas of concern. And regulating the abuse or corruption of the system is naturally critical in achieving security, safety, and privacy goals.

In this chapter, each of these areas has been addressed through various system protections and protocols. Specifically, a-ways through their enclosed structures can provide enhanced safety and security to travelers and items being transported in addition to more efficient emergency services. Autonomous transportation systems can also effectively regulate deliveries between senders and recipients while providing better security and privacy, and they can regulate access to premises by autonomous vehicles carrying individuals. Likewise, autonomous doors offer additional measures to enhance security, safety, and privacy while also providing great flexibility to accommodate individual preferences. And these same structures can provide monitoring and surveillance for added protections as well. Comparing these opportunities to current transportation and delivery systems, it is quite clear that the proposed autonomous transportation system offers many advantages in meeting security and privacy needs.

15
Political Challenges

IN 2015, A total of 195 nations throughout the world signed the Paris Accord in an effort to show a collective focus to combat climate change and implement strategies for environmental protections. The U.S. agreed to target a reduction in greenhouse gas emissions by 26 percent or more by 2025 domestically in addition to providing $3 billion to poor countries by 2020 to support their efforts in these areas as well. The Accord was viewed as a significant international accomplishment for sustainability and clean energy. But within two years, the U.S. announced its intentions to back out of this agreement in favor of isolationist policies to protect domestic needs, specifically those related to coal, oil, and automotive industries. Despite the intuitive logic of the Paris Accord, political resistance to change prevailed resulting in the decision to withdraw the U.S.'s participation in the initiative.[102]

Political resistance and challenges can certainly be substantial, and often times, this is most notable when the pressure for major change exists. Change requires a leap of faith in many cases, and it also requires the ability to identify fears, evaluate their validity,

---

102 Shear, Michael D. "Trump Will Withdraw U.S. From Paris Climate Agreement." The New York Times, 2017. Retrieved from https://www.nytimes.com/2017/06/01/climate/trump-paris-climate-agreement.html?mcubz=3

and develop strategies that embrace the change while minimizing unwanted effects. In some cases, political pressure encourages the change, but in others, resistance is present simply because change reflects the unknown. Pursuing clean energy and new transportation systems means existing structures will be disrupted, and the potential effect this might have on jobs and economies provokes worry and concern. Rather than embracing the change, it becomes easier to simply resist and stick with the status quo.

Though it seems the U.S. has developed an "ostrich approach" to climate change at the moment by sticking its head in the sand, political pressures to change are just as powerful as those that oppose it when it comes to clean energy and transportation. In this chapter, we will explore these political forces from the grass roots level all the way to global pressures for reform. As a result, you will appreciate how a change to a cleaner, more efficient transportation system is inevitable if we are to meet the nation's needs. Fortunately, the autonomous transportation model proposed allows just that. Change can start locally, for example inside buildings, and progress in stages. As opposed to cars which required roads to spread everywhere to be useful.

Political Inertia and Preserving the Status Quo

At the time, this book is being written, twenty-nine states in the U.S. currently have laws that allow medical use of marijuana. This includes 8 states that also allow recreational use of marijuana. Despite this, however, marijuana remains a Schedule I drug according to the federal Drug Enforcement Agency (DEA), and it remains a banned substance for use in any capacity.[103] How can this be? How can the government at a national level ignore the fact that over half the states have recognized the medicinal properties of this substance? Not only

---

[103] Governing.com. "State Marijuana Laws in 2017 Map." Governing.com Website, 2017. Retrieved from http://www.governing.com/gov-data/state-marijuana-laws-map-medical-recreational.html

is it much less costly compared to synthetic drugs, but it is also a very effective treatment for pain and cancer-related nausea, and the alternative is opioids which are a national fatality epidemic. At face value, the federal government's position doesn't seem logical.

In considering federal policies related to marijuana legalization, various pressures can be evaluated. Economic factors, job and employment effects, research evidence, and social opinions can each influence the ultimate policies that evolve.[104] Interestingly, each of these areas seem to support legalization of this substance, yet some resistance still exists. Why? Two reasons. One relates to the fact that more pressing policy issues currently take precedence over legalization of marijuana. The other involves an inherent resistance to change from existing businesses and social ideals. In this regard, political resistance to an autonomous transportation system is very much the same.

In considering autonomous transportation, we have already shown how such a system will be safer, more efficient, and more secure when compared to current transportation models. Likewise, the system proposed offers advantages in health, sustainability and in environmental protections, and it provides tremendous economic advantages at the same time. The main reason for political resistance has little to do with these areas as a result. Instead, an immediate push for such a system from a national perspective is not a high priority (even though it should be!). Instead, border protection, domestic job preservation, foreign trade agreements, and healthcare reform receive much more attention. The squeaky wheel tends to get the grease.

At the same time, major players have strong investments in maintaining the status quo. As previously discussed, the oil industry and

---

104 Chemerinsky, Erwin. "Why legalizing marijuana will be much harder than you think." The Washington Post, 2016. Retrieved from https://www.washingtonpost.com/news/in-theory/wp/2016/04/27/why-legalizing-marijuana-is-much-harder-than-you-think/?utm_term=.cfe1dc918299

automotive manufacturers would like things to stay just the way they are. They enjoy a competitive advantage under the current transportation system, and therefore, billions of dollars are spent by these industries for lobbyists to influence government policy in their favor. And at the same time, social complacency and conservative attitudes against change delay progress as well. Despite the obvious inconveniences, costs, and detrimental effects imposed on us by our current transportation system, these issues are not pressing enough on a day-to-day basis to force a concerted effort socially to insist on change.

While these factors have delayed progress in adopting technologies that would lead to an autonomous transportation model, the winds of change are upon us. Ultimately, a tipping point occurs that forces change despite areas of political resistance, and that tipping point is getting close. The demand for change is coming from many different areas today, and these are converging to place considerable pressure on policymakers to rethink current transportation and energy systems.

The Political Winds of Change

Understanding the innate tendency to avoid change and to cling to what is familiar, at some point change occurs whether we want it to or not. In many cases, a change does not come from a single source of influence but instead from a confluence of areas. This is the current situation when it comes to the adoption of better transportation models. Pressures for change are individual choices. Understanding the specific developments in each of these areas can provide a better picture of how political resistance will eventually succumb to the need for a better way.

As you are aware, the autonomous transportation system proposed utilizes clean energy alternatives, and therefore, the pressure for better energy systems is directly related to the need for better transportation systems as well. As nations and communities see the need for more sustainable energy solutions, transportation sectors will naturally be areas where significant gains can be made toward these objectives. This is already evident in the commitment for all major car companies to develop electric cars. While this pursuit is consumer-driven, changes in social preferences and ideals related to sustainable energy and environmental protections are also promoting these decisions.

Let's take a look internationally for evidence of these trends. Did you know that the life expectancy in parts of China has declined by over 5 years recently as a result of smog and air pollution? This is a tremendous figure in such a short period of time, and pollution from automobiles naturally plays a significant role. As a result, Chinese citizens are using face masks and air filters to combat the ill health effects, but this is an inadequate solution to a serious problem so political pressures have driven the Chinese to make massive investments in sustainable energy and to curtail coal burning.[105] For example, more battery electric vehicles were sold in China in 2016 than the next six nations combined.[106] Similar problems are developing in India as well as a result of rapid industrialization and the increased use of conventional cars. Eventually, these countries will be forced to seek alternative energy solutions and better transportation models, and this in turn will place greater pressure on the U.S. to do the same. Even if the US national policies do not push the car companies to build more efficient and electric cars, the reality of demands from China, which is now the largest car market in the world, and other countries are forcing them to build electric cars.

From a national perspective, domestic political pressures are also present. Despite powerful lobbyist groups promoting preservation of the status quo, various factors will demand alternative energy use and transportation models. For example, health care costs continue to rise, and with an aging population, resources must be used to

---

[105] Hook, Leslie. "China Smog Cuts 5.5 Years From Average Life Expectancy, Study Finds." CNBC.com, 2013. Retrieved from https://www.cnbc.com/id/100871380

[106] Jones, Mark. "How the US and China compare on action against climate change." World Economic Forum, 2017. Retrieved from https://www.weforum.org/agenda/2017/06/how-china-and-us-compare-on-climate-action/

provide these services. Given the safety hazards with current transportation systems (car crashes, pollution, etc.), and the marked inefficiencies and costs associated with them, models that reduce health risks and substantially reduce expenditures will have dual benefits in addressing healthcare needs of the country.

Unfortunately, many recent policies at the national level are incongruent with the pursuit of alternative energy solutions and more advanced transportation systems. Agendas that seek to save the coal industry and put autoworkers back to work are not only costlier but also less efficient and sustainable compared to policies adopting new technologies. It's no wonder state and local governments are taking matters into their own hands. As noted in the recent movie *An Inconvenient Sequel*, Mayor Dale Ross' decision to move his city of Georgetown, Texas, toward total reliance on wind and solar energy solutions and away from oil and coal is simply one of logic and common sense. Being the "reddest town in the reddest county in Texas," adopting change that boosts the economy is critical.[107]

Other examples of this exist from non-economic perspectives as well. California has been a long-time leader of alternative energy development. Certainly, these projects make sense from an economical point of view, but social preferences in the state have been progressive for decades encouraging sustainability policies. Most recently, in his comments supporting the rationale for backing out of the Paris Accord, President Donald Trump stated that he was elected to represent the citizens of Pittsburgh, not Paris. But in response to this statement, the mayor of Pittsburgh commented that the city would continue to support all the policies of the Paris Accord for the community, their economy and their future.[108] Ten states and many cities have committed to the Paris climate accord. The US is already

---

107 Wilkinson, Alissa. "Al Gore's new Inconvenient Truth sequel is a strange artifact of a post-truth year." Vox.com, 2017. Retrieved from https://www.vox.com/culture/2017/7/28/16035852/an-inconvenient-sequel-review-post-truth

108 Shear, 2017.

halfway to achieving its commitment for 2025, and current plans, with the commitments of states, cities, and companies, appear likely to meet the remaining half.[109] Even if national politics are ignoring social trends and political pressures, states and towns are not.

Finally, changes at the community level and among private enterprises pose additional pressures for political change toward cleaner and more efficient transportation systems. We have discussed how Tesla has turned the automobile industry upside down with its innovative strategies for electric (and sexy) car transportation. Similarly, large corporations like Apple and Google have dedicated their resources toward establishing complexes that embrace renewable energy while also exploring autonomous transportation. And presently, the solar industry employs twice as many workers as the coal, oil and gas industries combined when it comes to the production of electricity.[110] When private enterprises see the logic and rationale of adopting these new technologies, especially companies that exert tremendous influence over the national economy and job markets, you can bet political change will soon follow.

Perhaps the most powerful seeds for change are planted when individuals and communities decide to move in a direction of needed change. Just a few years ago, it would have been a novelty to see an electric car or a solar roof. But today, these are rather common along with other necessary infrastructures like electric car recharging stations. Marked increases in recycling behaviors have occurred showing social commitments to the environment and sustainability. And the rise of ride-hailing services to reduce costs and improve

---

109 Tabuchi, Hiroko, and Henry Fountain. "Bucking Trump, These Cities, States and Companies Commit to Paris Accord." The New York Times, 2017. Retrieved from https://www.nytimes.com/2017/06/01/climate/american-cities-climate-standards.html

110 McCarthy, Niall. "Solar Employs More People in U.S. Electricity Generation Than Oil, Coal and Gas Combined." Forbes.com, 2017. Retrieved from https://www.forbes.com/sites/niallmccarthy/2017/01/25/u-s-solar-energy-employs-more-people-than-oil-coal-and-gas-combined-infographic/#2c0523c62800

transportation convenience shows a need for change as well. Once these trends become powerful enough, and the threat of losing one's constituency becomes strong enough, policymakers will then be convinced to embrace change.

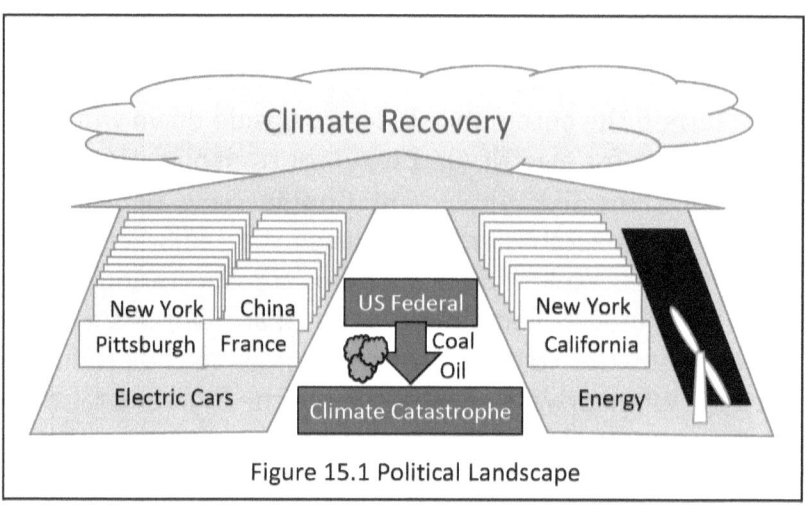

Figure 15.1 Political Landscape

A natural resistance to change often exists at many levels of society. Social views and values can change slowly, and those in power currently prefer the status quo to the unknown of a different approach. But eventually, a tipping point is reached resulting in major changes. While this tipping point may result from a single shift in technology, it often comes from several different sources. In addition to economic factors, health and social imperatives also play a significant role in promoting change. As discussed in this chapter, such sources for change in energy and transportation include international, national, state and local areas. These include not only public-sector influences but also those involving the private sector as well.

When it comes to autonomous transportation systems, the tipping point is near. Because transportation is closely linked to energy, pressures to adopt clean energy solutions will naturally affect decisions about the future of transportation systems. But the political pressures for change extend well beyond energy alone. Issues related to transportation efficiency, safety, security, and convenience will also progressively influence policy decisions at every level. And because these changes make logical sense from economic and social perspectives, political support will follow. It is not a matter of if, but a matter when…and the "when" is closer than you think.

16
Intentional Communities

DO YOU LIKE the community you live in? Does your neighborhood meet your needs? If you are like most people, you decided where you would live based on a variety of factors. Naturally, the area's cost of living and the affordability of your house were factors, but at the same time, you likely considered your home's proximity to your place of employment as well. Particularly in larger cities, there is a trade-off between commute times and price tags. If you work in the heart of the city, a shorter commute time usually means a higher priced house.

These are not the only factors that likely determined where you live. If you have children, school system quality was probably a major consideration. Likewise, basic community services and the aesthetics of the area influenced your decision. After considering all of these various qualities, you then devised some criteria that helped you make the final determination about your place of residence. But despite all of this information, one thing was missing. You had no idea what your neighbors would be like. What types of personalities would you encounter? What would their interests be? What type of lifestyle would they lead? While neighborhoods offer some ability to generalize about these answers, specifics are almost always lacking.

Given this scenario, your decision on where to live is constrained by two key factors…geography and information access. Let's consider geography first. Because your ability to move from one place to another quickly and affordably is limited in most cities, where you work and want to "play" are major factors to consider when selecting a place to live. And if you want to work and play in popular urban areas, you compete with others, which naturally drives up costs. Because our current transportation system is inefficient in many ways, our decision on where to live is geographically constrained because we can only live so many miles from where we work. This becomes one of the more significant determining factors in our ultimate choice rather than specifics about a community.

We are less constrained by poor information access; however, it still limits our ability to determine our ideal place to live. Real estate and chamber of commerce websites have a great deal of information about schools, shopping centers, crime rates, community services, and more. But they do not have any information about the interests, lifestyles, and preferences of individual neighborhoods and neighbors within these communities. This is not to suggest privacy rights should be ignored in order to get this information. But if such information was made available through voluntary efforts by individuals themselves to create better communities, this could greatly impact our ability to select the best place to live to meet our own needs.

With a transportation system that is faster, more efficient, cheaper, and less intrusive to our environment, we are able to relax many of the geographic constraints just mentioned. Instead of having to live within a certain proximity to our place of work, options for our residence expand significantly. At the same time, opportunities to access better educational resources, healthcare, and other areas of interest expand tremendously since the autonomous transportation system can get you there easily and quickly. Without these constraints, you can now choose where you live based on a completely different set of criteria. What if those criteria were focused on the actual community in which you were going to live…its people, their interests, their shared goals?

This in essence is an "intentional community." An intentional community is one where a group of people form a community based on personal characteristics of its members rather than on the physical criteria of the area. Without the geographic constraints imposed by an inefficient and costly transportation system, intentional communities become a possibility, and in the process, opportunities for enhanced creativity, social supports, innovation, and better quality of life become a reality. In this chapter, we will explore the concept of intentional communities and how autonomous transportation can provide the key infrastructure for such advances.

Intentionally Choosing Your Ideal Community

Based on 2016 expenditure reports, American consumers spent 16 percent of their budget on transportation and 33 percent on housing.[111] Thus, on average, half of our expenditures are consumed by these two categories. With the autonomous transportation system described, however, the costs for transportation fall dramatically, and the ability to access areas at distances farther away becomes possible. Likewise, with the development of linear cities, cost of building houses drops substantially for everyone. Combine these benefits with the additional time we gain through autonomous transportation efficiency, we will not only have more economic resources available but time resources as well.

With this being understood, the options where you might live increase significantly. What variables would you now consider if you could choose to live anywhere? Would you live in a community with people who shared your passion for athletic activities? Perhaps, you would choose a neighborhood where the group shared a common focus for continuing education. Maybe, if you had a health condition or physical needs, you might select a community that offered greater mutual support systems. Or you might live in a community that

---

111 Bureau of Labor Statistics. "Consumer expenditures – 2016." BLS Website, 2017. Retrieved from https://www.bls.gov/news.release/pdf/cesan.pdf

enjoyed a lifestyle centered on culinary passions. The key factor here is that a better transportation system enables us to consider new factors in determining the community in which we live, and these new factors have great potential in enhancing our lives, individually and collectively. Likewise, as linear cities develop, businesses and industries will be literally "downstairs" allowing less travel anyway. Thus, it would be expected that intentional communities focused on a select group of interests would likely be in close proximity to associated businesses and industries within their linear city.

What was the neighborhood like in which you grew up? For some of us, the experience was very positive while for others it was not. Warren grew up in a neighborhood that had no children his age within walking distance, so his early childhood was somewhat isolated outside of school. This had a major impact on his development, but fortunately his parents were terrific. Others may have grown up in neighborhoods where parents of friends served as key role models and sources of inspiration. This would similarly have a great impact on development and outcomes.

Regardless of your experience, you can appreciate the influence your immediate community had on you during those critical years of development. Even as an adult, your surrounding community affects your life in many ways, and this often becomes even more important as we age. Being able to choose where you live based on the key features of a community rather than on practical economic and geographic factors alone invites amazing new opportunities. Given the opportunity, most will likely embrace this chance to be surrounded by people who share common interests, compatible goals, and supportive lifestyles.

Let's consider some of the more exciting opportunities that intentional communities might offer. Consider a young, single mother who has a child with a severe food sensitivity or allergy. In addition to needing support to assist with childcare, she could also benefit from a community dedicated to investigating the cause and treatment options for her child's ailments. In addition, she spends a significant amount of her time outside of her job researching and preparing food her child can eat. In this regard, she has much to offer her community. For this woman, her quality of life might be greatly enhanced by living in a community focused on these health issues, and at the same time, she could enrich the community greatly with her capacity to share knowledge and information, not to mention good food, with others.

Here is another example. An older couple has decided to retire and choose a new community for themselves. Having been educators all their life, they are happy to retire but would like to continue to teach others as a hobby. With this being a priority for them, the couple seeks a community where education and learning is a shared interest, and likewise, a community where child and adult tutoring opportunities are available. While they gain tremendous benefit out of helping others learn and succeed, they also gain opportunities to partake in the community's educational workshops on a regular basis. Similarly, the presence of people like these would attract families with children to benefit from this education opportunity.

The ability to choose a community in which to live "intentionally" may focus on a variety of things. In addition to shared interests and hobbies, these decisions may be based on collective pursuits, health and other personal needs, lifestyle compatibility, and more. The obvious benefit is that people are allowed to better match their wants and needs with a community that is able to provide these things. But at the same time, communities, and society as a whole, benefit by allowing a variety of human activities and pursuits to contribute to the collective good. Many of us spend hours of our personal time learning new information and acquiring skills and talents that are never utilized at work or in any other organized setting. Intentional communities, however, provide settings where these untapped resources can be shared.

While all the previously described factors identify key motivations for intentional communities to form, the most powerful one will most likely be economic in nature. For centuries, people have migrated to city centers for obvious reasons...greater opportunity for success and happiness. With intentional communities, untapped resources will be better utilized, and at the same time, creativity and productivity will be enhanced in these intentional communities. As a result, the creations within intentional communities will be attractive to other intentional communities, and to others as well. In this way, economic incentives will also exist as these ideas, creations and new processes are shared. And as economic incentives develop within these communities, incentives for people to join them will increase as well. The autonomous transportation system makes it easier to share the innovative products of these communities. Thus, while the initial decision to join an intentional community will be most prominent among motivated and inspired individuals, in time, economic drivers will promote their growth among others as well.

Matching People to Form Intentional Communities
Whenever the word "match" comes up today, many people immediately think about online dating services. With this in mind, matching individuals and their families to intentional communities will *not* compare to these services. After all, we are not just talking about an isolated "bad date" encounter. We are evaluating where you will live for an extended period of time if not for the rest of your life. Given this fact, the matching process between people and their communities will be quite involved and thorough. The process will not only be extended over a period of time, but it will also involve participation by the individuals and interaction with the potential community members in advance before making final choices.

This may sound like a significant investment of time and energy. It is! But we are investing in something very important...our happiness, our health, and a better society. In addition, the autonomous transportation system facilitates these interactions ahead of time by being efficient, accessible, and affordable. While preliminary interactions with a community may be online and virtual, the autonomous transportation system will provide in-person interactions readily without the actual travel demands compromising our personal time to any significant degree. In this way, you will be able to have a rich experience with a potential community's members before making any final decisions about forming and joining a community.

So, where do we start when trying to match individuals and their intentional communities? The first step is to define your goals. This may sound simple, but in actuality, most of us are not adept at defining our goals, especially when it comes to where we might want to live. You may have broad goals, like leading a healthy lifestyle. Or you may have very specific goals, like becoming a top-level executive in a Fortune 500 company before the age of 35 years. The matching system will provide extensive help in defining and refining your goals. Regardless of the goals you select, these would be matched with other people that would support these goals. Likewise, the role you would play in the community as well as your learning objectives would be considered.

In addition to identifying goals, roles, and learning objectives, the matching process would also include a number of activities and interactions between you and your potential community members. These would help you and the community determine compatibility and allow you to be an active participant in the planning of the community. This is important for a couple of reasons. The more involved you are with the community in planning its functions and structures, the more invested you will be in its success. At the same time, this participation allows you to better assess whether the individuals' and community's goals are complementary to your own (and likewise, the community would be doing the same).

While every community will have its own characteristics, there are common features. For a community to be successful, it needs people with a variety of different characteristics, depending on the overall goals. This might include starters, organizers, cheerleaders, researchers, thinkers, detailers, and others. This matching and planning process will include a number of training programs both to fill in the skills you need to meet your goals and to be a valued member of the intentional community. Also, these training programs would help you understand and build ties with your future community members. Or they might even help determine whether the community and you are a good fit. Creativity and problem solving of all forms benefit from different perspectives, so you shouldn't be surprised if your ideal intentional community has some members that appear surprising at first glance. However, as you get into the training and planning process, the contributions of each member will become apparent with an understanding why such diversity is essential.

Warren: Let me share a story about a friend and colleague of mine. For many years, my friend has offered a one-week course to help "rewire" your brain, so I decided to enroll. Over the week, numerous activities were required (with necessary supports) for me to complete. Some tasks were things I never thought I could do (like write a poem or compose song lyrics). Others offered opportunities for me to see different perspectives (from vantage points of being a talker, listener or observer). In any case, after the week was over, I was amazed how my thought processes had changed. From being faster at word jumbles to the use of colors and diagrams to energize my work activities to enhanced sight reading of piano music, my thought processes had significantly changed for the better. And that was after only one week!

This story is provided to highlight the opportunities for training people to be more successful members in their intentional communities. Because we don't have widely publicized examples of these types of communities, it may be hard to imagine how such a community might work most effectively or how you can best participate in such an environment. In this regard, training would be quite important. And at the same time, training highlights the importance of continuous learning that these intentional communities would naturally support. Regardless of the shared interests and objectives, each community would strive to grow and advance through progressive understanding and higher levels of knowledge whether this might be centered on health, learning, a specific lifestyle, or a number of other interests. All of this is facilitated simply because of the freedoms a more efficient and autonomous transportation system provide.

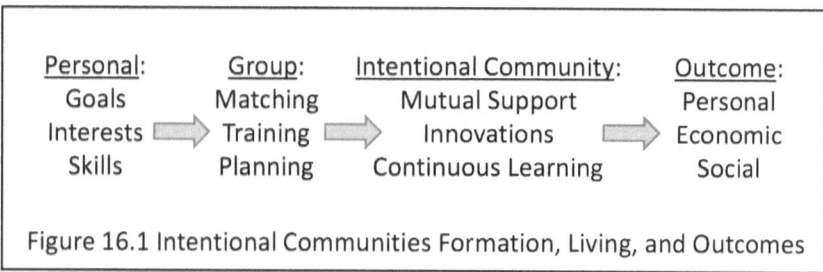

Figure 16.1 Intentional Communities Formation, Living, and Outcomes

Life in an Intentional Community

At least initially, these Intentional Communities will tend to attract highly motivated and inquisitive people, so their communities will be centers of excellence in specific fields. For example, if you are a researcher of a particular disease, allergy, or other difficult health condition, you will likely want to be where many people with that condition reside, where you can gather detailed information over extended periods, and conduct your research as opposed to simply seeing them once in a while. Likewise, this would thus be a great place to be if you have that disease or condition for access to the latest information. These may also need additional supports, such as special diets, so these needs will attract people interested in supporting this community of individuals. A combination of vocation, avocation, and social service opportunities will create vibrant intentional communities. These will then serve both as models and advertisements for the whole sustainable community model.

While an efficient and affordable transportation system provides the necessary infrastructure for the intentional communities described, these communities will not happen overnight. As noted, planning among members of a community will be needed once compatible goals are defined. But such planning will be able to enjoy an abundance of creativity and innovation in the process. How things get done, the physical layout of the community, which resources are shared, and many other aspects of the community will be determined by its members in an active, collaborative atmosphere. Therefore, the exact details of an intentional community in terms of how it operates and looks will vary based on these inputs and decisions.

Of course, not everyone will want to participate in these sustainable-type communities. Early adopters will likely be visionaries and enthusiasts, and this will facilitate creative solutions in how early intentional communities will look and function, and thus improve the experience for later intentional communities. In some cases, individuals may prefer to be on their own outside of these communities, and that is certainly an option. And in some instances, individuals may not function well in these communities despite wanting to belong. All of these challenges will need to be continually addressed, but this by itself offers new opportunities. For instance, intentional communities may develop around common interests to improve the matching process or to improve member satisfaction within communities. While the exact appearance of these communities is difficult to predict, the ability to choose where you live based on those around you, instead of simply by location and cost factors, will drive their development and progress.

So, what might life be like in an intentional community? For one, after you have participated in the planning of the community's functions and structures, you would enjoy significant support in pursuing your goals while mutually supporting others. In addition, the community would likely thrive collectively allowing sharing of specific expertise and resources. For example, key neighbors may be adept at providing emotional and psychological support during times of stress for you. Some members may want to engage in reciprocal learning activities where each member shares their knowledge and experiences in a variety of areas. And you will likely be surrounded by a number of mentors and role models related to your own personal goals.

In addition to these potential features of intentional communities, the opportunity for greater productivity also exists. As previously mentioned, intentional communities allow unique skills, talents, interests, and knowledge of its members to be shared in constructive ways so that everyone in the community benefits. But more importantly, intentional communities place individuals with compatible interests and goals in close proximity to one another. This is incredibly important when it comes to productivity and achieving goals. In my personal experience at Bell Labs, research there indicated that half of all productive interactions among individuals were unplanned. Likewise, for every 100 feet of separation between individuals, joint research productivity tends to drop 50 percent. By placing people with shared interests and compatible goals together, the chance for success increases significantly.

To illustrate the importance of proximity and interactions, consider the old Bell Labs research headquarters in Murray Hill, New Jersey. This was the home of inventions like the transistor, lasers, and information theory, to name just a few. The building had long hallways with offices on either side and a common cafeteria and library. On their way in to work in the morning, to the cafeteria at lunch, and home at night, each person walked by offices and labs and ran into people in the halls. These daily interactions were well known for contributing to the extraordinary innovations from this one building. When it comes to productivity, proximity matters. This is yet another area where intentional communities have the potential to flourish.

Life in an intentional community is likely to be many things to many people. Sustainable communities have the ability to provide social needs when more formal structures are inadequate. They have the ability to better utilize our natural resources to the benefit of everyone. Likewise, intentional communities enhance the capacity for productivity and social advancement. Intentional communities have the ability to excel in major areas like healthcare, learning, lifestyle and urban planning. And they contribute substantially to the quality and standards of life for us as individuals as well as for society as a whole. With an efficient, sustainable, affordable and autonomous transportation system in place, these types of communities become realistic possibilities for a much brighter future. The creativity and energy engendered by intentional communities are ideal for inventing, developing, and building the many new technologies and systems for autonomous transportation, linear cities and sustainability.

17
Living Spaces

IF YOU THINK about how we choose to live, some of our decisions seem a bit ironic. For example, many luxury homes have more bathrooms than they do people, and many homes go unused for months at a time. All the while, we continue to heat and cool these unused spaces. In the past, we have lived in suburbs far away from our jobs and colleagues just to have spacious neighborhoods and homes in which to live, but in the process, we sacrificed time and energy commuting back and forth to work. And today, many people are moving back to urban centers only to find that the prices have soared, forcing longtime residents to move to the suburbs, and that many of the residential areas cannot provide basic needs, such as food and other necessities.

Some of the most recent trends toward urban living can be tied to our need to be in close proximity to one another. Why? Primarily because proximity leads to availability of services, as well as creativity and productivity. In fact, as an urban population doubles, the level of productivity and wealth increases 15 percent above what would be expected.[112] In contrast, suburban isolation has the opposite effect on productivity as do the isolating effects of social media and

---

[112] Florida, Richard. The Rise of the Creative Class--Revisited: Revised and Expanded. Basic Books (AZ), 2014.

electronic communications. We need to be close to one another in order to tap into our creative potential, and this is a big reason that intentional communities will organically evolve as new transportation systems facilitate travel. We will want to be close to those with whom we share common interests and goals, and this sharing effect has the potential to influence how we live as well.

At first thought, it may not seem obvious how transportation affects the spaces in which we choose to live. However, transportation affects not only where we decide to live but how we live. Interstates, highways, and automobiles had a major impact on suburban living trends as did commuter rails and air travel. As an autonomous transportation system is implemented, this will have similarly major effects on the living spaces we prefer. This will not only involve new physical considerations but similarly cultural shifts in our thinking as well. In this chapter, we will explore some concepts for living space designs enabled by an autonomous transportation system.

Transportation and Living

We have already talked about how an autonomous transportation system can impact the communities in which we live. Because productivity and creativity are facilitated by being close to others with shared interests and goals, intentional communities will develop around specific ideas and concepts. Our choice to live in these communities will be encouraged by supports we receive from the communities as well as from economic incentives. So, in this way, transportation will influence decisions about where we will want to live in general. Because we can travel farther, faster, and easier, our options of which intentional community to live in will be greatly increased.

Understanding the impact autonomous transportation systems can have on intentional community development, we can also appreciate how linear cities will affect our living decisions. While a-ways and autonomous elevators provide the means by which we can access a variety of industrial, commercial, and community-related areas, we will naturally want to live closest to those areas that we plan to frequent the most. This not only includes our places of work, but it also pertains to people in our intentional communities with whom we collaborate and create. Though we can now travel to remote areas for learning, healthcare, recreation and more, we will still want to be near the communities that share our interests and goals. Why? Because that will allow us to best realize our creative and productive potential — for example, Lloyds of London originated in a coffeehouse frequented by ship captains.

How likely is this to actually happen? Social trends are already supporting these notions. You may not even realize it, but the growth of coffee shops in urban areas relate in part to the ability of individuals to meet and congregate in a central location more easily. Certainly, the quality of coffee plays a role, but coffee shops offer much more than good coffee. These are essentially locations where smaller versions of intentional communities can exist and co-create. In addition, these spaces offer tremendous flexibility and mobility. Without being confined to a specific office location, you can meet your colleagues in a variety of locations based on convenience and preference.

This raises another important point. In addition to providing flexibility and convenience, coffee shops also promote the concept of sharing space. This concept has been expanded into other business models today based on the success of coffee shops. For example, most major cities now have companies that rent office space and business services to individuals and small companies on a temporary basis. These clients may only need an actual physical office space a couple of days a month. As a result, they rent a single office for that amount of time while sharing the entire space with dozens of other people. It's a win-win…they reduce their office-related expenses while enjoying the ability to interact with others enhancing their productivity and creativity.

The same concept applies to mixed-use high-rise buildings in urban centers. While they have residences on higher levels, lower levels will often have residential common areas, business offices, and even retail stores. In addition to promoting greater occupancy, these mixed-use buildings allow residents to have immediate access to a variety of resources they desire. If you simply consider how autonomous transportation will promote the same thing through more effective and efficient transportation, you can readily see how this will impact our choices about where we will live, work, and play.

Current social trends favor a fundamental shift in our preferences about living spaces. Because we are progressively pursuing greater mobility, productivity, and creativity, we are much more open to the idea of sharing spaces with others. So far, this has applied mostly to office spaces and multi-use buildings, but the trend is clearly moving in this direction. With the addition of an autonomous transportation system, and with the development of linear cities and intentional communities, these preferences will soon spill over into residential spaces as well.

Shared Spaces and Walliture

A tremendous potential exists when it comes to sharing spaces, and an effective transportation system lies at the heart of this potential. You are likely aware of the impact ride-hailing services have had on travel today. These services are not only less costly, but often they are more efficient, and drivers get to recover some of their car expenses by "sharing" their car with others. But additional services by these types of businesses are expanding the benefits of sharing even further. For example, UberEATS uses its drivers to deliver food from a variety of restaurants to customers in their homes, offices, or any other place they like.[113]

If you think about this conceptually, UberEATS allows separation between a kitchen where the food is prepared and the place where the food is enjoyed. In other words, you can choose to have a meal from a wonderful restaurant in whatever ambience you desire. Now,

---

113 UberEATS. "How UberEATS works." Website, 2017. Retrieved from https://about.ubereats.com/

consider how this might be more effectively accomplished by autonomous vehicles. Not only would the delivery be autonomous and faster, but it would arrive in reusable containers specifically designed to keep your meal the perfect temperature. The potential this offers is surprising. Several people could meet in one location with each requesting their meal from different chefs, and through the autonomous transportation system, each meal would arrive at the same time perfectly ready to eat. This has obvious advantages for individuals with food allergies and specific food preferences.

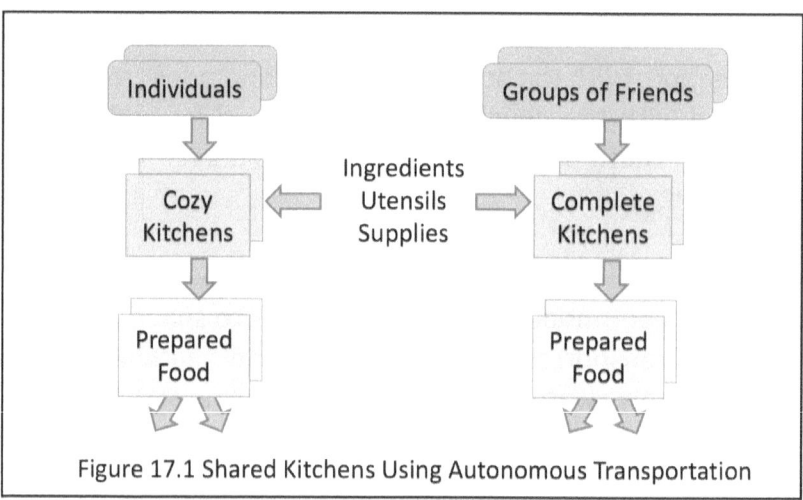

Figure 17.1 Shared Kitchens Using Autonomous Transportation

As you can see, autonomous vehicles can greatly facilitate sharing by providing the ability to deliver items to different locations. But autonomous vehicles have the capacity to promote sharing in other ways. In many warehouses, event spaces, and conference rooms today, accordion-like partitions exist to allow flexibility in the organization of spaces for different uses. Chairs, tables and other furniture can be added, removed and/or arranged to suit different functions. But in each of these scenarios, a fair amount of time and human labor is needed to make space transformations happen. But what if autonomous vehicles could perform these functions instead? These space transformations could occur autonomously, even including cleaning, allowing greater speed and flexibility in space utilization.

While these activities show how we could take advantage of new opportunities for sharing spaces, these ideas are only scratching the surface. If you have ever visited an Ikea furniture showroom, you might have noticed how you move through the store seemingly from "room to room." But the rooms created by Ikea are pieces of furniture on rollers, which allows Ikea personnel the ability to recreate spaces within the store over and over again with minimal inconvenience. With this in mind, what if we designed furniture in our own living spaces in a similar fashion? If so, we could recreate the room or space we needed at a specific time in the same common area. Instead of having hallways, foyers, guest bedrooms and formal dining areas taking up valuable, but rarely-used space, these spaces could be reconfigured to meet our current needs, and this would allow us to markedly reduce the total amount of space required.

You might think this is a little far-fetched, but Murphy beds have been in existence for many decades as a means to accomplish this very same task by folding a bed up onto a wall. Expanding these ideas further, we could apply this not only to specific pieces of furniture but to entire walls as well. Walls could be moved around to create new spaces, and they could have built-in components to accommodate specific needs, such as tables, stove tops, and sofas. Likewise, when they are not in use, walls could be compressed together to allow more open space for other activities. Just like some filing systems have walls of drawers that can be condensed when particular records are not being accessed, the walls within our homes could be designed in similar fashions. This system of mobile pieces of furniture and wall units can be called "walliture."

The incentives to adopt walliture and flexible living spaces would be several. Economic drivers will certainly exist. Less space means reduced expenses related to heating and cooling. Likewise, individuals as well as businesses might choose to share their space with others to reduce costs even further. In fact, some intentional communities may elect to have several shared kitchen facilities for the entire community rather than each individual living space having an elaborate kitchen. Meal sharing might be encouraged this way allowing those who like cooking to provide meals to others who do not enjoy this activity in return for some type of sharing. Cooking can be lonely, so shared kitchens would allow people to cook together, and some enterprising people who like to cook may turn remote delivery into a business, without the hassle of opening a restaurant.

As discussed previously, intentional communities allow untapped resources to be utilized by the community by providing a structure and system through which they can be appreciated. The sharing of living space among the intentional community furthers these efforts. Some members of the community will have specific talents and interests (such as cooking) that could benefit other members who lack this interest or talent. Shared space could thus be dedicated for these activities allowing better customization of individual spaces based on preference while still enjoying benefits for the community. And it is certainly likely that some intentional communities will concentrate on these specific areas in designing more effective walliture and shared space designs. Building walliture to meet the particular needs of a community is an example of creativity with local payoff, and potential for broader markets as well.

At this point, we have been describing how individual living spaces and those of the intentional community can benefit from space sharing and walliture. But the same concepts can be applied to commercial and industrial entities also. Have you ever noticed how much space goes unused throughout the day in a restaurant in between dining times? What about movie theaters and conference centers? By using walliture and shared space concepts, these spaces could host a variety of activities throughout the day. For example, a space may offer yoga classes in the morning, a dining area at meal times, a business meeting room in the afternoon, and a move theatre in the evening. The space that would be saved in the process, as well as the cost reductions

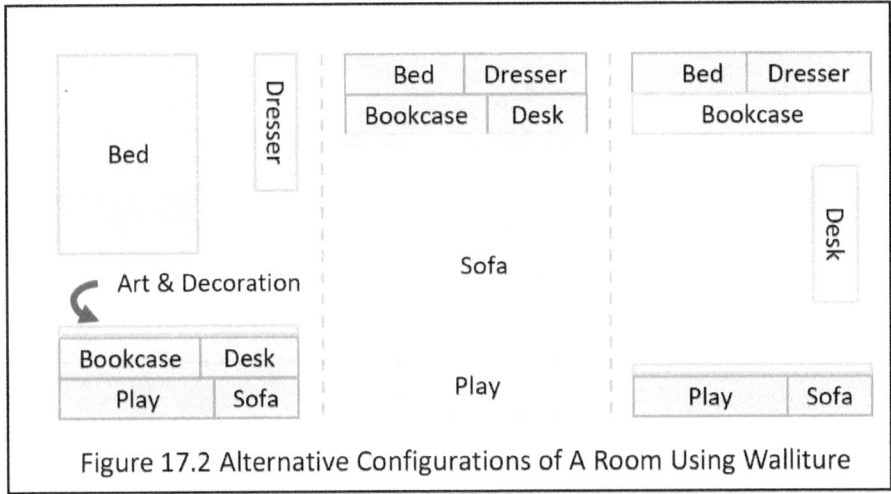

Figure 17.2 Alternative Configurations of A Room Using Walliture

experienced by each business, would be significant. This also contributes to continuous activity making the community seem alive rather than having numerous closed facilities contributing to a deserted feel.

The opportunities for more efficient and desirable living spaces would be notably increased when intentional communities and linear cities are considered. Likewise, concepts like walliture and shared spaces represent one possibility that could make this a reality. All of this becomes possible when an autonomous transportation system is in place offering enhanced speed, efficiency, and ease of travel. While other concepts may turn out to be excellent alternatives to these latter ideas, the fact remains that a better transportation system lies at the heart of these improved standards of living. Once such a system is in place, the opportunities for numerous other options and choices regarding the spaces in which we live, work, and socialize expand dramatically.

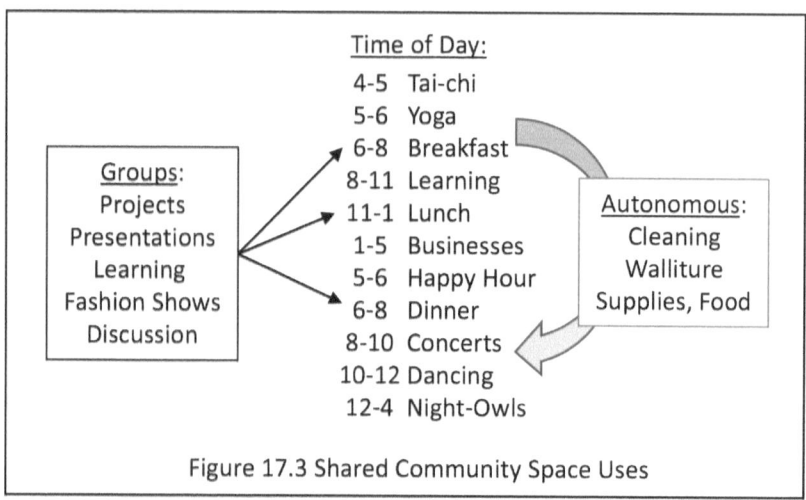

Figure 17.3 Shared Community Space Uses

Additional Considerations and Opportunities

The concepts presented in this chapter describing the link between autonomous transportation and advances in the living spaces provide a glimpse of what might evolve over time. But these are certainly not the only possibilities. The rise of businesses like Airbnb, VRBO (Vacation Rental By Owner), and others demonstrates how many people are sharing their living spaces with others when they are not occupying their properties. The changes proposed here regarding living spaces would facilitate these activities as well. For example, if you were away on vacation for a week or longer, you could arrange for others to use your space simply by condensing your private walliture into a much smaller area. Your intentional community might decide to organize this in a very structured way, or you may simply manage this on your own. Hot desking allows employees on different shifts to use the same desk, so the same approach could be applied to living space.

Likewise, society is much more mobile today than we were a couple of decades ago. Mobile devices and the Internet have promoted greater mobility, and this has affected how we live as well. Many people routinely relocate every few years to new areas, or they may travel often living in several different cities per year. The shared space and walliture concepts presented would make it much easier for individuals, families, and businesses to be more mobile. Entire intentional communities might maintain multiple locations, or even migrate. If they choose to retain a living space in one location while being in another, that space could be made available for use by others while they were away without compromising their private items. And with autonomous vehicles being able to easily move items to different locations (including walliture), moving from one place to another would be a piece of cake. And if you get tired of a particular piece of walliture, you can sell it or swap it for another piece while enjoying a change in your living space without the hassles of construction.

In mentioning autonomous vehicles from the perspective of living spaces, these could be used to further enhance living spaces and the quality of living. For instance, spatial reconfigurations could be done autonomously without requiring any physical labor at all. Autonomous vehicles could also adjust lighting within our living spaces to match our current needs, and the current spatial floor plan at a given time. All of this reduces costs allowing us to live in multiple locations if we so desire at different times while the existing space continues to be utilized in an efficient and productive manner.

As mentioned in our discussion about intentional communities, not everyone will want to share spaces or adopt new concepts like walliture. This is perfectly fine. Change often occurs through a series of gradual transitions, and these initial shifts regarding the spaces in which we live will be adopted by those who are more motivated to realize their potential. However, in time, incentives will increase and favor greater adoption of these changes by more and more people as they become refined. Some of these incentives will be economic in nature while others will simply be out of convenience. And without question, other creative solutions will be developed along the way. Regardless, the fact remains that an enhanced transportation system opens the door to some very exciting improvements in the spaces in which we live.

In short, the autonomous transportation system proposed is intimately linked to how we live. By enjoying greater proximity to others who share common interests, passions, and pursuits, productivity and creativity are enhanced. Likewise, the system places us in close proximity to related industries, businesses, and resources making our pursuits not only more efficient but also more likely to be successfully realized. And at the same time, we lower costs while advancing standards of living in the process. With an appreciation of this potential, you can see just how important and impactful an efficient and effective transportation system can be.

18
Learning Opportunities

BASED ON THE latest global education assessments of students, the Organization for Economic Co-operation and Development (OECD) has identified the U.S. as ranking around average for the 35 countries assessed in 2015. In fact, the U.S. actually ranked below average in mathematics and closer to average in science and reading.[114] While this may come as a surprise to you, it shouldn't. For several years now, the U.S. has progressively lost ground to other nations in educational success of its youth. Countries like Finland, Singapore and Peru out-perform the U.S. in all of these areas, and this has notable implications for the future.[115]

Naturally, educators and politicians are aware of these findings, and this has led to many educational reforms and programs seeking to correct a system that is suboptimal at best. But the educational system reflects only one of the major problems facing our society. As described in previous chapters of this book, technology displacements will occur (and in fact are already occurring), and this means we need a systematic way to retrain and educate adults as well as children. Another important issue pertains to the half-life of information today, which is amazingly short in many cases. You can no longer

---

114 OECD. "PISA 2015." Website, 2016. Retrieved from http://www.oecd.org/pisa/
115 Ibid.

rely on textbooks and lessons once learned in high school or college in performing the tasks at hand years later. Continuous learning is an absolute must for everyone moving forward. These are real issues that must be addressed as a society.

So, what do these issues related to education and learning have to do with transportation? Actually, a lot. You can certainly appreciate the impact the Internet and online learning programs have had on education systems and curriculums. A person can access tremendous amounts of information from the privacy of their bedroom or "on the fly" from their mobile device. Multimedia technologies can be interactive, but still have a long way to go to equal the best dynamic learning experiences. However, an efficient, autonomous transportation system offers new opportunities that allow not only greater access to new learning environments but also enhanced interpersonal interactions that would otherwise be impossible.

In this chapter, we will present a few noteworthy concepts about how the proposed autonomous transportation system can greatly enhance learning systems in this country. These suggestions are by no means meant to be a comprehensive portrayal of how education and learning systems need to be revamped and restructured, but they do identify many key problems and a few solutions related to our current systems. In doing so, we hope to call attention to exciting opportunities that can be created by a better transportation system, and at the same time, highlight how such a system supports the betterment of society in many ways outside of basic transportation improvements.

A Snapshot of Today's Learning Systems and Problems

Many years ago, some public schools, particularly in rural areas, had single classroom learning environments. Children of different ages would learn mathematics, reading, history, and science together. As the number of students grew, such learning environments rarely exist anymore, replaced by age-specific learning goals leading to segregation of classrooms and curriculums. This segregation was imposed based on student age and subject (rather than actual performance or level of interest), and as a result, we now have our current educational settings of age-specific grades and course classrooms.

While the single-room schoolhouse did have its shortcomings, one of the important advantages it offered was the diverse mix of students learning together. Older or more advanced students would take on the role of mentor to younger or less advanced students. By teaching the material, they became more knowledgeable. Likewise, learning was more dynamic and engaging simply because the material would be presented in different ways by different individuals, matching interests and previous knowledge. These types of learning environments are actually much more creative and effective, and as a result, these types of settings promote greater skills in problem solving, innovation, decision making, and multi-tasking. In most schools today, such environments remain uncommon.

This "warehousing" of students by age and grade has unfortunately encouraged a homogenous style of learning. In other words, educational programs are created with specific goals in mind based on age and grade. If these goals are not attained, students must persevere in these curriculums until the goal is achieved. Likewise, teachers, who are now held accountable for student performance, have been incentivized to "teach to the test" rather than striving to create a passion for learning in general. Combine these issues with a system laden with excess administrative burdens, inadequate teacher compensation, and below average learning outcomes, it is clear change is needed.

These are not the only issues with learning systems today. Several decades ago, many of the gaps in childhood education could be managed more effectively outside school settings. The presence of extended family members (and only one parent working instead of two) facilitated additional learning opportunities for children at home. But as we have transitioned to both parents working, single parent families, and as increasing mobility has occurred, access to extended families is becoming more and more limited. Combine this with the rapid advancement of information and knowledge, it is easy to appreciate that these prior supports can no longer effectively address shortcomings of formal education systems.

Some noteworthy research highlights some of these points. Did you know that soon after a child is old enough to start hearing words, they have the ability to learn between 10 to 20 words a week? Over the course of a year, this allows an incredible increase in vocabulary. However, there is one major caveat...they have to be exposed to an array of new words on a regular basis. In comparing children from advantaged homes to those from disadvantaged ones, a significant discrepancy exists as early as pre-kindergarten. In fact, advantaged children's vocabulary is twice as large on average compared to disadvantaged children. More importantly, this percentage persists throughout their school years.[116] Without effective learning systems, and without extended families to bridge gaps, we are doing a disservice to millions of individuals as well as society at large.

For these reasons, effective education systems are very important, and as already mentioned, ongoing learning needs to occur throughout life based on the rapid expansion of information and knowledge today. New learning systems need to be developed that address these issues for both children and adults. This not only supports social advancements and strong communities, but it also supports innovation, creativity, and productivity. And most importantly, learning is something that should be both innate and satisfying. New learning systems need to embrace an E^3 model...they need to be effective, efficient as well as enjoyable.

---

[116] Biemiller, Andrew. "Teaching vocabulary." American Federation of Teachers, 2001. Retrieved from https://www.aft.org/periodical/american-educator/spring-2001/teaching-vocabulary

Open Learning Resource Networks

In 2001, the Wikimedia Foundation launched Wikipedia, a multilingual, open source content encyclopedia. The shared information portal allows anyone to provide and/or edit knowledge and information about a subject thus creating an extensive resource for learning. While the site may often lack credibility and reliability in particular content areas, it is without a doubt a powerful tool for nearly 400 million people monthly who want to learn about specific areas of interest. The online encyclopedia reports more than 41,000 active contributors to over 41 million web pages in 294 languages at present.[117]

The success of Wikipedia is not surprising. We have moved from an industrialized society to an information-based one, and the Internet has allowed us to access, share, and distribute information efficiently and effectively. Wikipedia simply facilitates this exchange in an open, easy-to-use format. But at the same time, you may wonder why thousands of individuals take the time to contribute information to these web pages without pay or compensation. In some cases, advantages related to self-promotion or marketing may exist, but for the most part, contributors gain little in return from an economic standpoint. It would seem these individuals simply enjoy the process of learning and teaching through a system that allows them to pursue their specific areas of interests.

With this in mind, one of the learning and educational systems that offers great potential for society and our communities is that of an *open learning resource network*. In these networks, a variety of resources are accessible to anyone...texts, articles, videos, multimedia, audio playbacks, etc. In addition to these "stored" learning resources, live events and activities are included. However, unlike current education models, learners would be able to set their own learning goals. Instead of having to conform to a particular curriculum, pace, and school of learning, learners of all ages could design their own learning system. This would inherently promote greater

---

117 Wikimedia Foundation. "About Wikipedia." Website, 2017. Retrieved from https://en.wikipedia.org/wiki/Wikipedia:About

enjoyment in learning and self-motivation. Likewise, better resource accessibility would facilitate efficacy and efficiency in learning.

Warren: A personal story highlights this point. When I was in school, I had a retired rear admiral as my algebra instructor. When it came time to teach us algebra, however, his methods were a bit unorthodox. He informed us that we would each determine the pace with which we completed the course text, and when each of us individually felt we had learned the chapter material, he would test us on it. In addition, he also rearranged student seating in the classroom on an ongoing basis based on who was most proficient in completing the algebra tests and coursework. In other words, those students who paced themselves faster got to sit closer to the front. For me, this motivated me to learn the material as quickly as I could, but it also provided me with a sense of control and autonomy over my education that was personally rewarding. By setting my own learning goals and schedule, the learning became more enjoyable — this learning experience was a turning point in my entire career. Not everyone in the class was similarly motivated, so after a few weeks the teacher began a more conventional approach with those students.

Open learning resource networks could greatly advance our opportunities in learning in a very robust way through information access and through greater self-control over the learning process. But you might be wondering what this has to do with an autonomous transportation system. This brings me to a second point about learning systems. In communities and societies where innovation, creativity, and productivity are valued, exposure to diversity is a highly coveted characteristic. This pertains to social interactions, and it pertains to the manner in which we learn. While an open learning resource network offers diversity and access to information, an autonomous transportation system expands this to an even greater degree through dynamic physical learning opportunities.

With our autonomous transportation system in place, learning environments can change multiple times throughout the day. For example, a learner may choose to begin their day at a Museum of

Mathematics program, proceed to a community-based project at noon, and then participate in a focused apprenticeship later in the day. Despite being a significant distance in between each learning setting, the transportation system allows safe and rapid transit between these areas while enabling the learner to review recently learned information or prepare for the next activity while in transit. In this way, learning becomes much more dynamic and interactive, and this promotes engagement and heightened interest for learning. At the same time, it allows learners to select the types of activities that align well with their own unique learning styles.

Experiences	Authentic Work
• Zoo, museum, planetarium	• Design a bridge
• Concert, stage play	• Help a master painter
• Rehearsal	• Record & edit video
• Historic site	• Act with a coach
• Farm	• Debug a computer program
• Book signing	• Sell shoes
• Behind the scenes tour	• Proofread
• Factory	• Mentor other learners

Figure 18.1 Sample Learning Resources

One of my friends and colleagues, Peter Kline, told me a story that highlights the importance of matching learning activities to learner interests. He was observing one of the worst performing middle schools in Chicago, and the teacher of the seventh-grade class had become frustrated with her ability to hold the students' attention and engage them in learning the three states of matter. Peter offered to help. He then asked the students to get up out of their chairs, to put their hands at their sides and press together, after a short period he asked them to safely move around slowly in a random manner, and finally he asked them to move more quickly. He had them sit down and asked them what they experienced. To the teacher's amazement, every hand shot up, the class became actively engaged, interested, and described all the features of the three states of matter in detail. By providing activities that better aligned with their learning styles, and excess energy, Peter accomplished the task effectively and efficiently.

Warren: Each of us learn through different methods. For me, I can learn more efficiently and effectively when I process visual or auditory information through note-taking and mental transformations and working one-on-one with a person knowledgeable in the area. A friend of mine actually learns best when intermittently dancing after reading specific information.

Some people learn best in isolation while others require social interaction. Failing to appreciate this can lead to ineffective learning environments with non-engaged learners. But with an open learning resource network and a transportation system that facilitates access to a variety of learning environments, these needs can be much better met. Based on a learner's past experiences the open learning network would recommend which learning resources would best help achieve their current learning goals. This takes advantage of the unique interests, skills, and knowledge of each learner, rather than trying to teach everyone the same thing at the same time in the same way. Matching is an essential component of a learning experience helping each learner achieve their best potential.

In contrast to an open learning resource network, today's education systems are much more limited. Curriculums and teaching methods are more constrained and less dynamic. Even an excellent teacher can only do one thing at a time. Likewise, assessments and goals are predetermined based on collective goals targeting conformity rather than individual goals that promote intrigue, interest and excitement. And even in settings where learners become engaged and excel, nuances in teaching styles, materials covered, and student activities are rarely collected and analyzed in a larger, systematic way to guide progress of the entire system. As a result, we tend to keep doing the same things over and over again without improving the overall education system.

This brings me to one last point concerning open learning resource networks. Just as other areas like healthcare and business are focusing on "big data" to provide key insights into progress and change for the future, learning systems have this same potential as well. Through effective open learning resource networks, massive amounts of data could be collected and analyzed about learners and resources, suitably anonymized. From a learner standpoint, both individual and collective information could be gained about learning styles, curriculums, learner settings, and learner activities and enjoyment. From a teacher perspective, this could provide information about which teaching styles, strategies, settings and activities are most effective, efficient, and enjoyable for each learner. From a learning resource perspective, improvements can be incorporated based on the characteristics of learners using a particular resource, and needs for new learning resources to serve particular learners and subjects are readily identified. This would provide necessary inputs that could jumpstart significant improvements in learning systems moving forward.

By facilitating enhanced access to diverse learning opportunities, an efficient, autonomous transportation system can greatly improve existing educational systems. Combining this with an open learning resource network that offers learner-centric options for learning, big data analytics, and assessment systems that invite learning rather than oppose it, significant advancements can be made. This not only applies to youth learning systems but adult learning models as well.

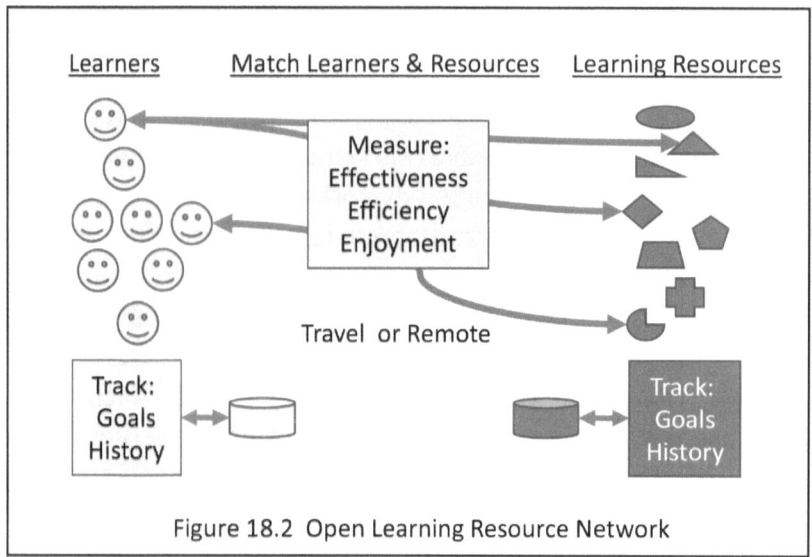

Figure 18.2 Open Learning Resource Network

Senior-Youth Matching Programs

For the first time in history, the world's elderly population is as large as the world's youth population under 5 years of age. Both groups now represent between 8 and 9 percent of the global population, and the percentage of individuals over age 65 years will soon reach double digit figures.[118] Of course, these statistics are not necessarily new information, and sectors related to healthcare and social services are scrambling to find solutions for the aging of our society. But as is the case in many instances, challenges can serve as potential opportunities for innovation and creativity. This is certainly the case when it comes to older adults and learning systems.

Let's examine how many older adults are treated today in our society. Many elderly individuals end up isolated, lonely, and bored. Families may have moved away, and forced retirement policies may have ended their careers before they were ready. And often, older

---

118 He, Wan, Daniel Goodkind, and Paul R. Kowal. *An aging world: 2015*. United States Census Bureau, 2016.

adults are discriminated against when it comes to other activities and jobs after retirement. Combine this with a culture that tends to glamorize youth and vibrancy, and you can appreciate how many seniors may feel. This occurs despite the wealth of information, knowledge, and wisdom they possess, and despite a true desire to be involved, connected, and a productive part of society.

Now, let's look at the flip side of the coin. As has already been mentioned, the ability of home environments to fill educational gaps of schools for children have diminished with the loss of the extended family. Yet, research shows that language stimulation at early ages has long lasting cognitive effects for children throughout their life.[119] Meeting this need could serve to significantly enhance learning. Likewise, many youths are in need of mentors, tutors, and companions throughout their formative years. If current school systems and families are unable to meet these needs, why couldn't other groups provide these services?

A senior-youth matching program would match older adults with youths. The knowledge and time that older adults have to offer would be shared with children and youth to help meet both specific learning objectives, but also broader needs such as human interactions and social supports. The program would include training for the adults, and monitoring to establish and meet goals. For example, a group of adults could meet prior to sessions with youth for training and support, and then meet with their youth partners in a common area for monitoring and assistance. Adults could meet afterwards to brainstorm and socialize. Just as intentional communities capture a variety of activities and resources within the community to advance productivity and innovation, senior-youth matching would provide the same benefits.

The initial assumption might be that the primary learning benefit would serve youth through such a program. But seniors would have the opportunity to enhance their learning and knowledge as well

---

[119] Biemiller, 2001.

through this arrangement. For example, adolescents could teach seniors the latest technologies and social trends to help them remain engaged and active within their communities. At the same time, ongoing learning helps ward off health conditions like Alzheimer's and other forms of dementia.[120] Clearly, a senior-youth matching program would be a win-win for all.

The beauty of such a program is its reciprocity, and it requires an organized system that allows seniors and youth to be well matched together at the right time. With today's digital technologies, the ability to develop such a system is certainly feasible, particularly in conjunction with an open learning network. And with an autonomous transportation system, barriers in mobility for both children and older adults can be easily overcome to enable such a program to be realized. The ramifications of a senior-youth matching program would be noteworthy not just in education and learning but in relation to better health outcomes as well. By having a sense of purpose and enhanced social engagement, seniors will experience a better quality of life in both physical and mental health areas. As the adults age, the support from the youth would increasingly flow to them. And at the same time, children will reap notable benefits socially and cognitively as well.

Because this type of program involves two potentially vulnerable segments of the population, appropriate screening and monitoring mechanisms would need to be in place. Regardless, with such protections in place, the potential advantages of a senior-youth matching program would greatly enhance and support ongoing efforts in education and learning. Current transportation systems are unable to provide the necessary structures for such a program to exist, but the autonomous transportation system proposed does. By providing greater mobility to these two population segments,

---

[120] Fratiglioni, Laura, Stephanie Paillard-Borg, and Bengt Winblad. "An active and socially integrated lifestyle in late life might protect against dementia." *The Lancet Neurology* 3, no. 6 (2004): 343-353.

these types of innovative ideas can be used to advance learning significantly.

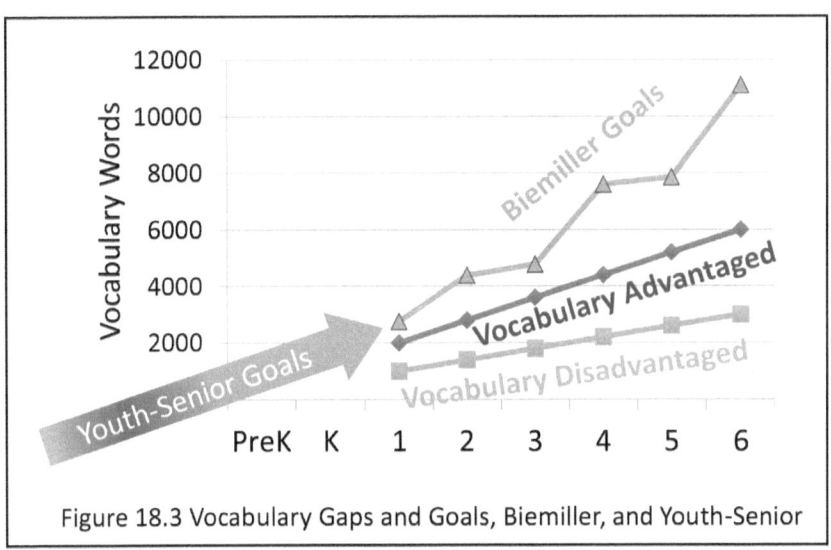

Figure 18.3 Vocabulary Gaps and Goals, Biemiller, and Youth-Senior

Life and Learning

When it comes to learning, much of what we learn is through experimentation. Research seeks to identify facts and knowledge by examining cause and effect situations in controlled environments. If the experiment reveals some hidden truth, and it can be validated by getting the same results over and over again, knowledge advances. In many ways, life is exactly the same way with the exception that a controlled environment is not possible. We move through a series of trials and errors that guides our behaviors. For us, we use the information gained to change future behaviors (feed-forward), and we share our experiences with others to help them as well (feedback). On a daily basis, this is how we learn.

With this in mind, learning does not stop upon completion of high school, college, graduate school, or trade school. Instead learning occurs throughout life for each of us regardless of whether the learning is formal and structured or not. However, what is not in place is some type of mechanism to capture the abundance of learning that is occurring constantly. The feed-forward and feedback information is rarely cataloged in any structured way to guide others, and methods of learning that are most effective are rarely examined in relation to our evolving education. And many of the most powerful resources that could help us learn are often beyond our grasp due to barriers to access — creating and using learning resources provides such a mechanism.

These problems associated with education and learning are likely to be resolved in the near future. Big data are just starting to enter the field of education, and data analytics will provide key insights about best practices in learning and education. At the same time, advances in information access will continue as will numerous types of media through which we can learn. But simultaneously, resource access and new learning opportunities will develop as better transportation removes mobility barriers. These systems not only allow us to gain better access to learning resources, but they also enhance a variety of social interactions that support learning further. From senior-youth programs, to apprenticeships, to intentional community projects, an autonomous transportation system is a key component that allows enhanced learning and education.

Learning is fun, satisfying, and a natural part of life regardless of whether you are young or old. Likewise, learning supports individual health, growth, and development while also facilitating the advancement of communities and societies. And with an aging population and declining education outcomes, social pressures will increasingly demand better solutions. The proposed autonomous transportation system and the implications for other technologies and practices create myriad business opportunities. These in turn create both the need and the opportunities for greatly expanded learning. With autonomous transportation in place, it will be much easier for these solutions to be developed and implemented.

19
Health Effects

THROUGHOUT THE BOOK, many beneficial health effects have been described in relation to autonomous transportation. Inherently, major reductions in transportation-related injuries and deaths should occur with such a system. In 2016 alone, over 40,000 deaths resulted from car crashes, and over 4.6 million injuries from car crashes required medical care. In total, these costs are estimated at $432 billion! And the most notable statistic is that 94 percent of all car crashes were related to human error.[121] Without question, our overall health as a nation will improve dramatically with an autonomous transportation system.

While these figures are substantial, autonomous transportation has the capacity to enhance health in other even more significant ways. From medication delivery to access to health-promoting activities and behaviors, enhanced transportation systems help us transition from a reactive model of healthcare to one that is proactive and preventative. In addition to having quality of life benefits, this also has notable cost-saving ramifications. Given that 17 percent of the

---

[121] National Safety Council. "NSC motor vehicle fatality estimates." Website, 2017. Retrieved from http://www.nsc.org/NewsDocuments/2017/12-month-estimates.pdf

nation's gross domestic product goes to healthcare expenses, and given that the population is aging rapidly, controlling costs is a significant area of interest.[122] Autonomous transportation can go a long way in helping with these efforts.

In this chapter, several additional health advantages that can be realized with autonomous transportation will be discussed. These insights are not meant to be comprehensive as many other potential health benefits exist with a more effective and efficient autonomous system, but major areas where immediate health improvements can be seen will be highlighted. With this overview, you will be able to appreciate the significant impact transportation can have on healthcare, health outcomes, and healthcare costs.

Advantages in Emergency Medical Management

When it comes to many medical conditions, time is of the essence. Every minute counts. The 9-1-1 system (in part) was established for these types of emergencies, and each year, roughly 12 million calls are placed to 9-1-1 for medical emergencies.[123] According to the American Heart Association, for individuals suffering from a cardiac arrest, the survival rate declines by 10 percent for every minute that care is delayed. For acute strokes, therapies must be administered within a few minutes to avoid permanent brain injury in many cases.[124] And cost estimates suggest that reducing emergency response times by a single minute could save over $7 billion in healthcare costs each year![125] Time is health, and health is money.

---

122 Centers for Medicare and Medicaid Services. "NHE fact sheet." Website, 2016. Retrieved from https://www.cms.gov/research-statistics-data-and-systems/statistics-trends-and-reports/nationalhealthexpenddata/nhe-fact-sheet.html

123 Rapid SOS. "Outcomes: Quantifying the impact of emergency response times." Website, 2015. Retrieved from https://www.cms.gov/research-statistics-data-and-systems/statistics-trends-and-reports/nationalhealthexpenddata/nhe-fact-sheet.html

124 Ibid.

125 Ibid.

With these figures in mind, the autonomous transportation system proposed can improve health outcomes and reduce healthcare costs substantially by the very nature of the systems that will be in place. Let's consider autonomous vehicles first. These vehicles will be equipped with emergency notification systems that will allow passengers to alert the system of a medical emergency. Likewise, the precise location of these vehicles will be known at all times, in addition to the closest medical care facilities. Therefore, the fastest route can be determined instantly, and the autonomous vehicle can transfer individuals to the appropriate care facility saving precious time when compared to today's transportation systems, rather than waiting for an ambulance to arrive and only then being transported to a facility.

In addition to the capabilities of autonomous vehicles, the overall transportation routing system will also play a role in reducing the time required to receive emergency care. Once an emergency has been identified, the system will invoke an emergency priority routing so that the autonomous vehicle can get to the desired location as fast as possible without other delays. If you have ever seen an ambulance or EMS vehicle trying to get to and from a car crash today, you can appreciate how such a routing system could reduce response times significantly. Here again, this will enhance health outcomes and reduce costs.

What if the person having a medical emergency is not in an autonomous vehicle or is too unstable to be transported? An autonomous transportation system can facilitate care here as well. Upon notification that a person needs immediate care, emergency medical services and/or equipment can be dispatched and routed to the person's exact location efficiently and without delays. Regardless whether the person is at home or in the community, these autonomous services would greatly improve access to emergency care by reducing response times. And in the process, regular transportation flows would proceed without significant interruptions. These benefits offer tremendous improvements to individual and population health when it comes to caring for those with emergency health conditions.

Advantages for Healthcare Access

It's no secret that limited healthcare access is a major issue in the U.S. Most of the challenges in accessing healthcare services have been described in relation to health insurance. However, many other issues related to access involve actual transportation barriers for many groups of people. The inability to use current transportation options, long distances to healthcare services, and reliance on others for transportation all reflect significant issues when it comes to healthcare access. And here again, an autonomous transportation system offers many benefits in relation to enhanced health.

Consider people living in rural areas. First of all, half of all car crashes occur in rural areas despite less than a third of all travel occurring here. Certainly, an autonomous transportation system would offer health advantages in this regard. But in addition, individuals living in rural areas have much farther to travel to reach specific healthcare services needed. This is complicated by the fact that fewer healthcare providers choose to provide services in these areas. This shortage of providers in rural areas persists despite the fact that many conditions like diabetes and heart disease are more prevalent among rural populations, and despite the fact that roughly 20 percent of American citizens live in these regions.[126] With an autonomous transportation in place that can rapidly and efficiently transport these individuals to the care services they need, access is notably enhanced, and health outcomes markedly improved for this segment of the population.

Other notable groups that suffer from limited healthcare access include people with disabilities and the elderly. Approximately one in

---

126 Rural Health Stanford School of Medicine. "Healthcare disparities and barriers to healthcare." Website, 2017. Retrieved from http://ruralhealth.stanford.edu/health-pros/factsheets/downloads/rural_fact_sheet_5.pdf

5 individuals have a disability, and as previously described, the elderly segment of the population is growing in number. Many of these individuals are unable to drive due to various conditions. These conditions include visual impairments, movement difficulties, a lack of coordination, or other functional limitations that make them dependent on others or various transportation services to access healthcare.[127] As a result, time delays in receiving care occur for many, and in other situations, appointments are missed or postponed. Combine these problems with the fact that these individuals commonly have a higher number of medical conditions, and it highlights how improved transportation models facilitate better health outcomes.

In addition to helping specific individuals and populations access healthcare services they need, autonomous transportation can help everyone access higher quality services. For example, suppose you had a rare genetic condition, and the closest specialist for your condition was located a few hundred miles away. Today, it would require a serious commitment of time, money, planning, and scheduling to see the specialist. In contrast, the autonomous transportation model we propose would allow you to accomplish this much faster and efficiently. You would likely receive better care more frequently, and you would suffer fewer inconveniences in the process. For anyone in need of special services, this system would greatly improve the ability to receive that care.

Advantages for a Healthier Lifestyle

For many decades, the U.S. healthcare system has focused on managing disease rather than pursuing disease prevention and health promotion. Despite clear evidence that a proactive system is more cost effective and yields better health quality, the system has been slow to change. But things are changing. Increasingly, reimbursement is

---

127 U.S. Census. "Nearly 1 in 5 people have a disability in the U.S. Census Bureau Reports." Website, 2012. Retrieved from https://www.census.gov/newsroom/releases/archives/miscellaneous/cb12-134.html

being linked to better patient outcomes rather than on the number of services provided. And this has significant implications in changing the approach to healthcare over time. Because of these changes, health promotion and encouragement for patients to lead healthier lifestyles will become a more important focus of healthcare services.

In this regard, the effects that an autonomous transportation system will have on diet, nutrition, and physical activity will be substantial. In terms of foods, the distribution, production, preparation, and consumption of various foods will certainly change. Growers and sellers of foodstuffs will be better linked to consumers without having to go through typical manufacturing and production companies and "middle men," and thus be much better compensated for their efforts than currently. Likewise, access to a higher variety of foods will be possible for larger populations because of improvements in transportation efficiency. These factors have the potential to drive down costs, expand dietary options, and promote healthier eating habits.

Let's consider a couple of examples to highlight these points. For an individual with food allergies or sensitivities, the ability to eat meals that are safe can be challenging. Many choose to cook at home in order to ensure they are not exposed to food items that could prompt ill health. As a result, their opportunities to dine out with friends are limited. Likewise, for many economically disadvantaged families, access to healthy foods may be limited and/or costly. This is often true in some urban environments, and this has contributed greatly to our obesity epidemic. Inexpensive, high-calorie, fast food becomes a necessary option for such families. In both of these situations, an autonomous transportation system could provide lower cost, safer, and healthier foods for consumption. And in turn, this would promote better health.

Other areas related to diet where transportation could offer advantages relate to food-borne illnesses. Outbreaks of E. Coli or other organisms in foods can often cause major health problems in large groups of people requiring urgent medical care. Food inspections are supposed to reduce the occurrence of these events, but inspections are naturally sporadic. What if this could be accomplished more routinely and without involving large numbers of inspection personnel? Interestingly, technologies are available that could allow this to happen in conjunction with an autonomous transportation system. "Lab on a chip" and rapid DNA sequence tests could be incorporated into transportation stations where autonomous vehicles carrying foodstuffs can detour for testing while en route. Sources of problems could be quickly identified and their products avoided. In addition to reducing the costs of inspections, such a system would expedite the process while providing better food screening.

Healthy lifestyles involve more than just diet and nutrition. Physical activity and exercise are other important health-related areas that prevent disease and promote health, and an autonomous transportation system is advantageous here as well. While individuals differ in terms of their preferences when it comes to exercise, two things typically promote greater participation. The first relates to opportunities to engage in social or group activities since this often makes exercise more enjoyable and fun. Let's face it...exercising alone can be boring, isolating, and cumbersome at times. By expanding the destinations where we can efficiently travel, we naturally open up an array of new group activities that may be attractive to us and help maintain our attention for much longer periods of time. As a result, we overcome some of the common barriers to exercise, with intentional communities being a major help.

The other way an autonomous transportation system can promote physical activity is through variety. With current travel limitations, we may be limited to the number of gyms, fitness groups, and outdoors activities available in our immediate vicinity. This all changes with a better transportation system that is faster, more efficient, and autonomous. We might choose to hike in the mountains one day and then go ocean kayaking the next despite the two locales being a hundred miles apart. Likewise, the expansion of green spaces, and transportation in enclosed a-ways, that would be possible with linear cities would similarly increase opportunities for the number of outdoor exercise options. And intentional communities would likely encourage these exercise programs through mutual goals and support to further promote engagement and participation.

Unlike many congested environments today where access to a healthy lifestyle is constrained, a new transportation system could markedly improve access and opportunity. Such a system has clear advantages when it comes to nutrition and diet, and it also encourages greater exercise by offering more attractive options for individuals to consider. Combine these features with health education and learning opportunities, and it becomes evident that these would have positive impacts on our health as individuals and as a society.

Advantages in Medication Safety and Efficacy

Medication errors represent a serious problem in the United States. In fact, more than 7,000 deaths occur each year as a result of some type of medication errors both in and out of the hospital, and this does not factor in the occurrence of illicit drug use and overdose.[128] In the U.S. alone, nearly half of the population takes at least one prescription medication, while nearly a quarter take 2 or more medications.[129] And each of these medications has the potential for side effects, adverse reactions, and drug interactions that undermine medical safety and quality patient healthcare.

In considering these statistics, any system that reduced medication errors would be welcomed. These issues not only cause health problems for individuals, but they also contribute substantially to healthcare costs. Every time someone has an adverse reaction to a

---

128 Donaldson, Molla S., Janet M. Corrigan, and Linda T. Kohn, eds. To err is human: building a safer health system. Vol. 6. National Academies Press, 2000.
129 Centers for Disease Control. "Therapeutic drug use." Website, 2017. Retrieved from https://www.cdc.gov/nchs/fastats/drug-use-therapeutic.htm

medication, symptoms often require additional medical attention. In some cases, this requires new treatments and care, and in others, disability or death may occur. Being able to ensure medications are taken at the right time, in the right amount, and for the right length of time can significantly help reduce these problems. This is where an autonomous transportation system and "just-in-time" medication delivery can have major impacts on healthcare.

Let's consider how such a system can help prevent medication problems. The first major area where just-in-time medications would help involves medication adherence. Medication adherence refers to taking medications as prescribed. This pertains not only to the right amount and timing of the dose, but it also relates to taking the full prescription as ordered. Errors in dosing, timing, and duration account for significant healthcare costs and undesirable outcomes. While poor adherence to dosage and timing recommendations can result in poorly treated health conditions and/or unwanted medication effects, poor adherence in relation to duration of treatment has led to major healthcare problems like antibiotic-resistant bacteria. These issues could be notably reduced if an autonomous transportation system could deliver the dose of medication you required at precisely the right time. Similarly, it would continue to do so until the full course of medication treatment was completed.

Just-in-time medication has clear benefits in safety and health, but at the same time, this system would also greatly reduce medication waste. How many times have you started a medication that required subsequent adjustments? Even worse, many times the first medication tried may not work, and you must switch the medication until the desired effects is achieved. Unfortunately, you have already purchased either a 30-day or 90-day supply since these are the standard amounts prescribed. In many cases, the excess medication you purchased and never used gets literally flushed down the toilet, causing problems downstream.

This unnecessary waste is completely avoided with autonomous medication delivery that is just-in-time. Excess medication does not have to be purchased as only the doses taken would be delivered and administered. If you needed to switch medications, this would be done without having to discard a stockpile of medication. Just-in-time delivery would also help reduce accidental poisonings at home. Many people hold onto past prescriptions that were never taken or finished in case they may have some need for these drugs at a later time. Rarely is this the case, and the stockpiled medicine cabinet serves as a potential source for accidental poisonings, especially for children. In fact, accidental poisonings were the number one cause of unintentional deaths in the U.S. for individuals between the ages of 25 and 64 years and were second only to car crashes for those ages 5 to 24 years.[130] An autonomous medication delivery system would significantly improve this situation.

The advantages just-in-time medication delivery would offer extend beyond reductions in excessive prescription dosing and poor compliance with medication instructions. This would also help deter prescription medication theft, loss, and illegal sales. These problems are particularly notable when dealing with controlled substances like opioids. When prescriptions for these medications allow access to quantities that last several weeks, the risk for illegal sales and theft increase. Likewise, risks for potential over-dosage and addiction climb as well in these situations. Limiting the number of dosages of these controlled substances to only the amount required at a specific time would reduce these risks dramatically. And by ensuring delivery to the right person, theft of these medications diminishes as well.

Because our health system predominantly focuses on the treatment of existing disease, a significant proportion of the population

---

130 Centers for Disease Control. "Ten leading causes of injury deaths by age group highlighting unintentional injury deaths, United States 2015." Website, 2015. Retrieved from https://www.cdc.gov/injury/images/lc-charts/leading_causes_of_injury_deaths_unintentional_injury_2015_1050w760h.gif

takes prescribed drugs. In fact, the global pharmaceutical market exceeded $1 trillion in revenues in 2014![131] Given the expenses associated with these drugs, their potential hazards, and the significant waste in the system, autonomous transportation and delivery of these medications through just-in-time programs can help improve healthcare greatly. Considering the healthcare crisis at hand in the U.S., such a system would not only resolve many existing cost problems associated with the healthcare system but likewise improve overall health of the population at large.

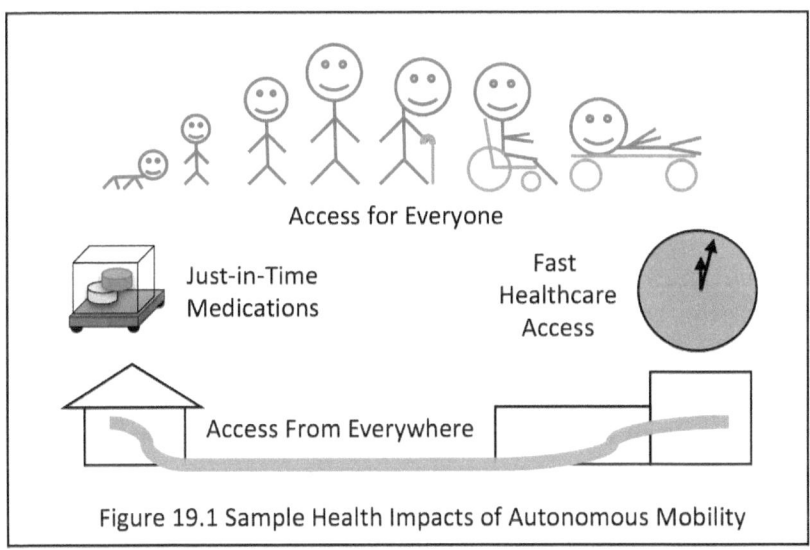

Figure 19.1 Sample Health Impacts of Autonomous Mobility

Additional Indirect Health Advantages

As you can see, an autonomous delivery system can offer many advantages to our nation's healthcare system and to our individual health overall. Better access, better healthcare, and enhanced patient safety would undoubtedly result with this transportation and delivery system. From emergency response times to medication delivery

---

131 Statistica. "Global pharmaceutical industry, statistics and facts." Website, 2015. Retrieved from https://www.statista.com/topics/1764/global-pharmaceutical-industry/

just-in-time, the magnitude of these positive effects would be substantial. But these are just the more obvious areas where health and healthcare would be improved with such a system.

In terms of indirect benefits on health, the autonomous transportation system proposed would also be helpful in promoting better environmental health. Substantial reductions in air pollution would reduce respiratory illnesses while promoting a better quality of life and longevity. And new opportunities for improved sanitation would likely evolve through the use of an autonomous transportation system. The areas described in this chapter highlight how the autonomous transportation system would provide immediate advantages to health, but in time, additional benefits would also be realized through other innovative and creative concepts.

The effects of an enhanced transportation system are thus quite extensive, and those involving health and healthcare portray the vast potential for improvement such a system allows. Using new solutions developed through intentional communities and through existing industries, the nation's healthcare woes can be better addressed using the foundation the autonomous transportation system provides. While the economics, social conveniences, and environmental pressures will more likely drive the change to an autonomous transportation model, the impact this model could have on sectors like healthcare further support why change is both inevitable and at hand. And health facilities may be some of the first to implement autonomous transportation.

20
Economic Drivers for an Inevitable Change

AS WE GO through our normal day, we often notice how things might be improved. Sometimes, the change that should be adopted seems so obvious, yet the change doesn't immediately occur. For example, the change to electric cars from those that run on oil and gasoline makes logical sense from many perspectives. They are less costly to operate, require less maintenance and repair, and they are naturally better for the environment and our own personal health. And these technologies have been available for well over a decade. If this is the case, then why doesn't the majority of Americans drive electric cars today?

The answer to this question is twofold. The first reason pertains to how we react to change. Inherently, there is a natural resistance based on what we have come to know as familiar and comfortable. A certain amount of persuasion must take place in order to overcome our own inertia and to take that leap forward into something new. As a result, we often wait until a crisis is present before such a change is embraced. In other words, when push finally comes to shove, we reluctantly accept that change is needed.

The second answer, and the reason push often comes to shove, is that economic drivers have not yet become strong enough to force the change. While protecting the environment and our health has long-term repercussions economically on us as individuals and as a community, we tend to pay more attention to those things that affect our pocketbooks in the moment. For this reason, real change often happens when the current costs of continuing to do the same old thing exceeds the cost of doing things a new way.

From these perspectives, we believe an array of economic drivers now command a major change in our transportation systems. A paradigm shift is imminent in this regard, and the winds of change are already in the air. By examining both the direct and indirect economic drivers behind this change, you will be better able to appreciate why an autonomous transportation system will be realized in the near future. From consumer benefits to widespread cost savings to society itself, the money that can be saved with such a system (and earned) will force us to finally shift gears from our current mode of travel to one that is inherently better from every aspect.

Direct Economic Drivers

Understanding that the squeaky wheel tends to get the grease, immediate and direct cost savings for consumers and society alike will help pave the way for an autonomous transportation system. Let's look at this from a consumer's point of view first. As has been described in various chapters of this book, autonomous vehicles will be substantially smaller in size, and at the same time, be much less complex in their design. Because of this, the costs to construct these vehicles (as well as the costs to maintain and/or repair them) will be significantly less for us as individuals. With fewer working parts, and with the possibility of 3D printers being able to construct these small vehicles within our own homes and free open source software, consumers can quickly make the move from one type of transportation to the other.

Consumers will enjoy additional economic benefits as well with the autonomous transportation system. Because the system is safer and protected from various environmental hazards and human operator errors, insurance rates will be notably less. Likewise, operating costs will be much less when compared to today's transportation. The electrical costs to operate an autonomous vehicle will be a fraction of operational expenses of today's gas vehicles. This is already a current incentive for many to select electric cars over traditional models today.

The ability to dramatically reduce direct transportation costs for individuals will shift buying patterns and consumer preferences very quickly. Think about how quickly ride hailing services advanced based on these same economic drivers. Similarly, the rapid rise of electric scooters and shared electric bikes demonstrates consumer desire for innovative transportation solutions. Is it any surprise these same types of economic drivers will hasten the race to autonomous transportation as well? These factors, along with new opportunities to share autonomous vehicles, will enable individuals to enjoy more efficient and safer transportation at a significantly lower cost. Based on conservative estimates, it is highly likely these direct costs may be as little as a tenth of what consumers pay today.

Consumers typically drive market changes, but at the same time, broader system costs also can do the same when it comes to transportation. Direct cost reductions to governments, industries, and businesses will help foster change and the adoption of an autonomous transportation system. As has been described in earlier chapters, many cities and states are facing serious financial issues in relation to infrastructure repairs and replacements. Basic economic analysis shows it will be more affordable to consider a completely new type of transportation infrastructure rather than rebuilding one that is decrepit and will clearly be obsolete in the near future. Once maintenance cost savings are factored into these analyses, many municipalities and state departments of transportation systems will opt for change.

Businesses and industries will also experience the same cost advantages with an autonomous transportation system as consumers and governments. In a competitive market, pressure to reduce expenses and maximize profits will require these organizations to continually consider new options. To date, companies like Amazon have done this through robotics, shared delivery systems, and other innovative programs. These cutting-edge companies will likely begin to explore more efficient autonomous transportation and delivery systems soon as well. Reduced costs for transportation mean either higher profits or a competitive advantage in their market (or both). Given the number of companies that experience high transportation and delivery costs, economic pressure will require them to consider new alternatives. And once the ball is rolling, other companies will soon follow as well.

Indirect Economic Drivers

Transportation affects so many aspects of our lives, and it should come as no surprise that enhanced transportation can benefit our bottom line in other areas. Indirect economic drivers exist for consumers and enterprises alike, and these can have similar influences on the decisions we make about transportation. Just as direct reductions in cost will promote adoption of an autonomous transportation system, so will these indirect effects. Understanding these economic drivers will further show you how autonomous transportation is within our immediate future.

We have already talked about the cost savings consumers will directly enjoy with the autonomous transportation system described, but in addition, other major areas will also drive change from an economic standpoint. The first area involves time. With autonomous transportation, we are free to do many other things while we travel rather than concentrate on driving the vehicle. The time it takes for us to get from one place to another will be markedly reduced. Combine this with the development of linear cities where many things we love are in close proximity to our homes, and you can begin to see how the opportunity to be more productive with our time will help drive change.

From this point of view, time offers you the potential for greater income if you choose to use the extra time for such pursuits. This might be accomplished individually if you so desire, or it might be pursued through your intentional community. The option to share your autonomous vehicle with others represents another indirect economic driver, or a direct driver if you charge for usage. And sharing your vehicle may evoke cultural shifts that encourage sharing of other things (including living spaces) in society. All of these are lifestyle features that reduce costs for individuals and families, and when considered as part of an autonomous transportation model, they will further support change and adoption of these new systems.

Health and education sectors will also indirectly support autonomous transportation systems as a result of reduced costs to consumers. From a healthcare perspective, these transportation models will facilitate access and better quality of care and help lower costs of overall healthcare. By eliminating medication waste, theft and misuse, pharmaceutical costs will be less as well, although improved access to better medications could go the other way, while still reducing overall health costs.

Education will follow a similar path. Not only will access to new learning resources and the quality of education improve, but youth and everyone will have better opportunities for economic success, and society will spend less on prisons and other expenses for failures in the current education system. While these indirect effects may not be the primary drivers of change, they will support and sustain momentum in that direction.

When considering indirect economic effects on other sectors of society, other notable factors will also play a role in supporting the change to an autonomous transportation system. One obvious one will involve opportunities for less expensive infrastructure costs. By housing other infrastructures within a-ways, durability and accessibility are improved. This has a major impact on lowering installation, maintenance and repair costs. These savings become even more significant when we start considering linear cities and intentional communities where resources and infrastructures are shared even more efficiently.

Businesses, industries and municipalities will all want to take advantage of these economic savings, and these benefits will help influence policy and strategy changes in the process. A new building could actually save money by including an autonomous way and autonomous infrastructure inside the building, making the building more valuable, and then getting others to share the costs by building on top. Thus, we could see new autonomous ways and infrastructure developing with private funding, rather than all being funded by the various levels of government. Health facilities and senior facilities are obvious initial opportunities, in addition to industries and businesses.

Lastly, the other important indirect economic driver for an autonomous transportation system relates to real estate markets and home construction. By incorporating autonomous transportation systems into our living communities and cities, the opportunity to recover land space will be significant. Because autonomous vehicles require less space, a-ways will be much smaller in size than corresponding roads. Likewise, by using nesting and continuous convoys, space requirements will be even less. And through the development of associated linear cities, both construction costs and land space demands will shrink even further. Ultimately, this positively affects the costs associated with housing as well as those associated with commercial spaces. Imagine highways being converted to linear cities because of the combination of savings and convenience, with surrounding open spaces.

It is important to realize that these indirect factors will affect many areas and be relevant to consumers as well as various enterprises.

Their role of influence will serve more as a catalyst once change begins to occur because of the many additional economic advantages a new transportation model offers. Some individuals and companies will be ambivalent about (or even resistant to) these changes, but as soon as these additional impacts are witnessed, they will soon follow, of necessity. Thus, in many ways, indirect economic drivers will be just as powerful in the long run in helping support the adoption of this type of transportation system.

Additional Economic Drivers

While cost reductions will perhaps be the more obvious economic drivers for change, revenue generating opportunities will also help support a new transportation model. These opportunities will come from a variety of areas involving both private and public sectors. Likewise, new unconventional enterprises will evolve offering new revenue streams. Because of these opportunities, the potential for economic gains will attract early adopters of these new technologies and systems, and once proven, additional interest will soon follow. In addition to the direct and indirect cost savings already described, these new areas will offer additional incomes to further support this change.

Let's consider some specific examples. With enhanced transportation, new business models will be created because traditional channels of production, distribution and delivery will no longer be the same. Imagine receiving fresh produce directly from a garden, or being able to enjoy a hot meal on a mountainside far from your favorite chef. Consider the possibility of receiving a design kit from an online vendor for the new autonomous vehicle you plan to create with your home 3D printer. The possibilities for new business opportunities for the autonomous transportation system described are essentially endless.

The same options for new enterprises and business pursuits will similarly exist for intentional communities. The degree of convenience, access, and proximity that the autonomous transportation system allows will promote collaboration, innovation and creativity among individuals who share common interests and passions. As a result, intentional communities will invest their energies in exciting new endeavors that will further enhance society. And the economic rewards when exchanging their innovations with others will help drive these endeavors. As described previously, the foundation of an enhanced transportation system provides the necessary infrastructure for these new opportunities.

Many of these additional economic drivers are difficult to visualize or predict since the implementation and realization of any new transportation system will be dynamic. But the potential for new business pursuits and innovation is compelling. Having an enhanced transportation system makes these things possible while also creating completely new opportunities for employment and career pursuits. These new revenue opportunities will catalyze the transformation of our current transportation system as we know it, and the ability to enjoy a much better transportation system will be a reality.

New Demand	Add Capacity
• Active Adult	• Airports
• Senior Facilities	• Hospitals
• Health Facilities	• Subways
• Warehouses	• Highways
• New Industries	• Sidewalks
• Personal needs	• Cities
• Hobbyists	• Housing

Cost Reduction	Unsafe Conditions
• Businesses	• Bridges
• Public Agencies	• Tunnels
• Cities	• Railroads
• Low-Income Housing	• Roads
• Defense	• Shore Communities

Figure 20.1 Why Build An Autonomous Transportation System?

Conclusion

Throughout the course of history, many visions of the future and its technologies have faded away never to be realized. Thus, it would be reasonable to ask why the concepts and models of an autonomous transportation system in this book might actually become a reality in our future. The most obvious reasons involve inherent economic drivers...after all, money talks. The freight train analysis shows an advantage of 30,400:1, which is so large that it can't be ignored. Providing a new transportation system that is faster, more efficient, autonomous, safer, cheaper, and more enjoyable will get everyone's attention. From individual consumers to businesses, communities, and governments, the financial incentives such a system brings to the table are substantial and not easily ignored. And with the potential for reducing costs and improving quality in areas like education, healthcare, and land use, many other economic benefits of this system will support radical change.

But economic drivers are not the only reason an autonomous transportation system is in our future. The availability of new technologies will also push us in this direction. Take a look around you. Developing countries are completely skipping wired communication technologies in favor of wireless ones. Without having to go through the costly and laborious development of wired infrastructure development, these nations are able to quickly jump into the 21st century in terms of digital communications. If you were a developing country, and autonomous technologies were available to you, less expensive to develop, and offered a markedly better transportation system, wouldn't you take that option over today's traditional transportation system?

Let's take things a bit further. Renewable energy resources are quickly becoming available for many regions of the world, and they are likewise being offered at prices comparable or less than fossil fuel resources. These opportunities are already disrupting the energy industry in major ways. For example, rooftop solar is a technology providing electricity directly to homes and offices in a much more efficient and climate-friendly way. This same technology offers tremendous opportunities for autonomous transportation as well by providing a-ways and autonomous vehicles with nearby energy resources.

In considering major changes throughout history, various crises often drive change as well. Crises during World War II stimulated the creation of several inventions and the adoption of new technologies. Likewise, poverty, famine and other threats have done the same. Today, we are facing an array of crises as a nation and as a global population. From a national perspective, education and health crises are major concerns as is our dependence of foreign energy resources. From a global point of view, climate change, water shortages, and poor sanitation systems reflect additional issues. Transportation systems are intimately related to these crises, and they can play a significant role in resolving these problems. And at the same time, they can directly contribute to improving energy, pollution, human safety and global health dilemmas.

Given these scenarios, we are at a unique and exciting confluence of events that greatly favor the adoption of autonomous technologies. The recipe includes amazing advances in communication technologies, information systems, data analytics, and renewable energy resources. In addition, it also involves multiple crises related to energy, the environment, health, and resource sustainability. An autonomous transportation system as proposed reflects a natural progression from current transportation models when these aspects are considered. Multiple drivers for these changes are already present, and it will soon be impossible to hold back these advances any longer.

In the early 1970s when Warren was at Bell Labs, it was often said that the fastest data transfer system involved putting a magnetic tape on a Greyhound bus. Today, magnetic tapes are a thing of the past, and it is quite probable Greyhound buses as we know them will be too. Advances in digital technologies disrupted the information world. We are now at the cusp of another disruptive shift. Autonomous technologies will soon change the physical world just as the Internet and digital technologies changed the information world. If you remember what it was like to suffer through slow dial-up modems, you can appreciate the dramatic changes that have occurred in the last couple of decades in terms of information access and communications. In the very near future, we will marvel at the same transformations involving transportation.

Today, if we are searching for the best loaf of bread available, we can find out a tremendous amount of information about the type of bread we desire. We can check out online reviews of various bakeries, read product descriptions and consumer opinions about specific loaves, and even search for the best prices. But when it comes to actually getting the loaf of bread, we still have to drive to the store or bakery, or we have to rely on some type of delivery system using human drivers to get it to us. We still rely on cars, trains, and trucks to move physical objects around…but this will soon change as autonomous technologies are employed.

Just as the Internet provided greater access, safety, and affordability to information, autonomous technologies will provide the same things for physical items. Using the autonomous transportation model described, we can have our loaf of bread autonomously delivered to us for a fraction of the cost without having to invest our time and energy driving to and from the store. At the same time, the system would be much safer, secure, and efficient. No pollution, no risk of human driver error, and no traffic delays. And for the third of the population who cannot currently drive, autonomous technologies provide great advantages for improving access to physical items and locations. Without question, these are very exciting times as we sit on the cusp of remarkable innovation and change.

So where might the adoption of autonomous transportation system first occur? Amazon has over 80,000 robots delivering shelves of goods, and over 15 million Roombas are vacuuming floors, so the transition has already started. Some of the most likely areas to get started involve contained systems or environments where prototypes can be tested and improved upon. Places like airports, senior living communities, healthcare complexes, and seaside communities with limited evacuation options and increased flooding might be areas where autonomous transportation models could be easily employed. Other areas are major enterprises involved in package delivery systems. Even urban commuter rail systems could be replaced with autonomous transportation models as part of initial implementation efforts, for example overlays on existing highways. The exact areas where these will first be employed are difficult to predict, but once employed, the proof that such systems are the wave of the future will likely accelerate their adoption in many other areas.

The seeds of change have already sprouted and taken firm root with electric cars, and self-driving cars are ready to leap off the drawing board. Do we anticipate that cars will be completely replaced? No. People today still ride and race horses, and many people still collect and drive antique cars. Self-driving cars will be an important part of the transition from today's sprawl to the future a-ways and linear cities. But what we do anticipate is that better forms of transportation will emerge quickly. Cities and suburbs, which were shaped by the automobile, will be transformed by new autonomous transportation systems. And the rest of our physical world as well as many other sectors will be affected also. What will our transportation systems look like in 20 years? The only safe bet is that they will be very different from today.

www.ingramcontent.com/pod-product-compliance
Lightning Source LLC
Chambersburg PA
CBHW021810170526
45157CB00007B/2529